Philip Stewart Robinson

The Poets and Nature

Reptiles, Fishes, and Insects

Philip Stewart Robinson

The Poets and Nature
Reptiles, Fishes, and Insects

ISBN/EAN: 9783337025885

Printed in Europe, USA, Canada, Australia, Japan

Cover: Foto ©berggeist007 / pixelio.de

More available books at **www.hansebooks.com**

THE
POETS AND NATURE

REPTILES, FISHES, AND INSECTS

BY

PHIL ROBINSON

AUTHOR OF "THE POETS' BIRDS," "THE POETS' BEASTS," ETC.

London
CHATTO & WINDUS, PICCADILLY
1893

PREFATORY NOTE.

In order to anticipate the discovery by others that I have omitted some of the Poet's Insects, I avail myself of the Author's privilege of a Preface to say that my list does not contain, except incidentally, any of the beetles (although such notable insects as the "book-worm" and the "death-watch" are of the tribe), any of the "gnats" or wasps, both so cordially detested by the Poets, or any of their favourite worms. Among the fishes, again, owing to a mishap to manuscript, there is a hiatus, which would be of no importance but for the consequent omission of Thomson's lines on Angling—already largely drawn upon, however, in my "Poets' Beasts."

Acting on the advice of my Friends (the Critics of my two previous volumes), I have added an Index to this, the third, and hope that their prediction of the increased usefulness of my work from such an addition may be fulfilled. The task which the publication of this volume completes was undertaken by the Author, and accepted by his Publishers as one which should be of use to men of letters; and it has been a great pleasure to notice, during the last four years, from the columns of newspapers and the pages of magazines, reviews, and books, that our expectations have been satisfactorily realised; and I can assure those Writers who, after availing themselves of the quotations which I had brought together expressly for their service, have gone on to reproduce my comments, that I accept their recognition of the object of my compilation as a compliment, even when they have translated me to their own books, a page at a time, without any acknowledgment, and even when, mistaking me for an obsolete and extinct Person, they have respectfully regretted that I had not survived to the Tennysonian Epoch, when many of my opinions might have been altered for the better.

Wherein Jean Ingelow is delightfully avenged. For having in my first volume premised that I was not dealing with " contemporary" Poets, I quoted largely from that most charming Writer ; and my feelings may or may not be imagined on receiving, a few days after the publication of my book, an invitation to a garden-party, signed " Jean Ingelow !" My pleasure in discovering my mistake was increased by being thus enabled to dedicate my second volume to one of the most exquisite Poets of Nature and most tender Interpreters of the speechless world.

PHIL ROBINSON.

January 1893.

CONTENTS.

PART I.

The Poets' Reptiles.

PART II.

The Poets' Fishes.

PART III.

The Poets' Insects.

PART I.
THE POETS' REPTILES.

THE POETS AND NATURE.

PART I.
THE POETS' REPTILES.

CHAPTER I.

"REPTILES AND VERMIN."

A REPTILE is not, perhaps, an amiable thing. Its name—
"that which creeps"—prejudices some of us against it.
Nor is there anything thoroughly unjustifiable in this. The
necessities of speech require a word that shall compendiously express the idea of the contemptible and crawling,
and at the same time the potentially hurtful. And "reptile"
fulfils this obnoxious duty. So when Beattie applies this
term of reproach to a servile poet, "the reptile muse, Swoln
from the sty, and rankling from the stews," or Byron to
a mean critic, they are not to be found fault with. The
sycophant in Shelley, the slave in Montgomery, even man—
"the poor reptile man and heir of woe" himself—in loftily-
moralising Greene, are metaphorically rendered, and not
unfairly, by a term that zoologically implies either a turtle,
a crocodile, a frog, a lizard, or a snake. Southey brings
some priests under the same category, and scattered up and
down in verse will be found scores of individuals whom the

poets, anxious to stigmatise as despicably base, denominate "reptiles."

Now all this is perfectly fair. We have attached to a certain word a certain metaphorical meaning, which is a very odious one. Bismarck called the Secret Service Vote "the Reptile Fund," and the Man of Iron includes in it all such miserable creatures as venal editors and spies. The self-seeking parasite, the insidious hypocrite, the cringing slave, deserve the worst we can say of them, and as we have decided that there is nothing worse to be said of such than "reptile"—reptiles let them be.

But here we must stop. Even the prerogatives of human beings do not extend further. They cannot outrage the sacred laws of justice even in the case of reptiles. So we have no right whatever to make the name of a particular thing mean what it does not, and then to transfer the arbitrary character which we have affixed to it back to the thing whose name we borrowed. We have absolutely no right whatever to begin by saying that sycophants, hypocrites, slaves, and assassins are "reptiles," and then to say that reptiles are sycophants, hypocrites, slaves, and assassins. Merely as a logical syllogism it is absurd and untenable. Here are the two premisses: Despicable men are reptiles; Reptiles are either turtles, crocodiles, lizards, frogs, or snakes. Work as you will with them, your conclusion must either be that no conclusion is possible, or else an absurd statement to the effect that a frog is a despicable man, or that sycophants, &c., are either turtles, crocodiles, &c.

But, setting logic aside, I contend that it is infinitely unjust to speak ill of an immense number of creatures, nearly all of which are either beautiful, directly useful to man, or harmless, simply because, in our usual high-handed way of dealing with the helpless, we have borrowed their collective names as figures of speech. Yet this is what most poets habitually do. Their toads are loathsome and their frogs obscene. Their chameleons are turncoats, and their scorpions traitors. Their snakes are utterly abominable.

Now I fail to see any justification for this. It strikes me as thoroughly immoral. Even snakes, against which human prejudice cites Scriptural authority, are admirable. They are one of the most splendid parables in all nature. Nothing that breathes less deserves the title of reptile—meaning by that word a despicable cowardly thing—than the creature that stands in Holy Writ itself as the semblance of a power that could defy Heaven and challenge terms with Omnipotence. I would even go further and venture to say that this, the poet's treatment of a large order of creatures, shows a deficiency of sympathy with nature which is not in accordance with the poetical tradition. For example, take the following from Montgomery :—

> " Reptiles were quickened into various birth,
> Loathsome, unsightly, swoln to obscene bulk,
> Lurk'd the dank toad beneath the infected turf;
> The slow-worm crawl'd, the light chameleon climb'd
> And changed his colour as his place he changed;
> The nimble lizard ran from bough to bough,
> Glancing through light, in shadow disappearing;
> The scorpion, many-eyed, with sting of fire,
> Bred there, the legion-fiend of creeping things."

But worse than this, as expressing a wider range of unsympathetic prejudice, are such sweeping lines as these of Coleridge :—

> " What if one reptile sting another reptile?
> Where is the crime? the goodly face of nature
> Hath one disfeaturing stain the less upon it."

The philosophy here is thoroughly bad-hearted and reprehensible.

Another poetical liberty which I consider only indifferently justified is to call insects "reptiles." Thus Thomson (as usual "shagged with horrors") addresses such pretty things as may-flies and butterflies as a "reptile throng," and it is worth noting how with his usual infelicity he speaks of these *reptiles* as being "winged, and by the light air upborne."

> "To sunny waters some
> By fatal instinct fly; where on the pool
> They, sportive, wheel; or, sailing down the stream,
> Are snatch'd immediate by the quick-eyed trout
> Or darting salmon. Through the greenwood glade
> Some love to stray; there lodged, amus'd, and fed
> In the fresh leaf. Luxurious, others make
> The meads their choice, and visit every flower,
> And every latent herb.
>
> Some to the house,
> The fold, and dairy, hungry, bend their flight,
> Sip round the pail, or taste the curdling cheese;
> Oft, inadvertent, from the milky stream
> They meet their fate; or, weltering in the bowl,
> With powerless wings around them wrapt, expire."

Wordsworth, again, calls the glow-worm "a very reptile," which is intolerable, seeing how he uses the word elsewhere. Eliza Cook, after her wont, speaks of cobwebs as

> "The bright slime that cunning reptiles spread
> To catch their prey."

But her use of the reptile idea is always thoroughly in character with her poetry generally. What can we say, for instance, of such a stanza as this where an unmasked villain is illustrated by a *skinned snake* :—

> "Why, why does Heaven bequeath such gifts
> To fascinate all eyes, that mark
> With magnet charm, till something lifts
> The mask, and shows how foully dark
> The dazzling reptile is within
> Beneath its painted, shining skin?"

But this lady's definition of reptiles is, like most ladies', very vague. They consider the word synonymous with "vermin," under which title they include all the creatures they most object to, such as rats, mice, spiders, black-beetles, earwigs, and snails.   The right of the sex to dislike what they choose is of course indisputable, and in the vary-ing technical definitions of the word "vermin," they have a plausible excuse for ranging far.   Thus, the professional "vermin-killer" is a rat-and-mole-catcher.

> "Som poison, that he might his ratouns quell
> And eke ther was a polkat in his nawe
> That, as he sayd, his capons had yslawe :
> And fayn he wolde him wrekin if he might
> Of vermin, that destroied hem by night."

On the gamekeeper's table of the proscribed are the weasel kind, and many birds, such as the owl, jay, hawk, heron, and hooded crow.   On the Continent, beasts of prey, such as wolves and foxes, are so styled.   In Australia official enactments call rabbits and wild horses "vermin."   In Western America the Red Indian himself goes under the same name.   And in the West Indies, according to Montgomery, the man "of colour"—

> " Lives there a reptile baser than the slave ?—
> Loathsome as death, corrupted as the grave,
> See the dull Creole."

Nor do poets of a more robust sort hesitate at similar licence.   Man himself, as in Cowper, is "vermin ;" lawyers in Somervile are "the vermin of debate ;" and courtiers in Thomson are "the vermin of state that on our substance feed."   Gutter-children have in Mackay the same ill-sounding name :—

> "Take them away !   Take them away
> Out of the gutter, the ooze, and slime,
> Where the little vermin paddle and crawl
> Till they grow and ripen into crime."

Criminals and "parasites" too, and Jesuits—and critics—are all "vermin."

Coming lower down, we find the polecat called vermin by Chaucer, the rat by many, the mole "vermin impotent and blind" by Butler, the woodlouse, spider, housefly, and a number of insects by others. So that "reptile" and "vermin" are virtually interchangeable terms with the poets. Each denotes or connotes the meanest individuals of every class of beings—those which the poets individually consider the meanest—but in either case they go so far wrong as to borrow a creature's name in order to convey an odious meaning, and then transfer the odium which they arbitrarily and capriciously attach to the word back to the creature.

A further curious complication in this high-handed confusion of terms is the use of the word "insect." It is employed as synonymous with reptile and vermin. Thus, man generally is an insect, and so too are special classes of men, notably faithless friends, courtiers, all kinds of sycophants and parasites, and pleasure-seekers generally—and riches.

> " The insect tribes of humankind,
> Each with its busy hum, or gilded wing,
> Its subtle webwork, or its venom'd sting."—*Rogers.*

> "All the vast stock of human progeny,
> Which now, like swarms of insects, crawl
> Upon the surface of earth's spacious ball,
> Must quit this hillock of mortality
> And in its bowels buried lie "—*Oldham.*

> "The swarm that in thy noontide beam were born,
> Gone to salute the rising morn."—*Gray.*

> "Ye tinsel insects whom a court maintains,
> That count your beauties only by your stains,
> Spin all your cobwebs o'er the eye of day;
> The Muse's wing shall brush you all away."—*Pope.*

> " The nameless insects of a court."—*Thomson.*

> " The pageant of a day, without one friend
>   To soothe his tortur'd mind ; all, all are fled,
>   For though they bask'd in his meridian ray,
>   The insects vanish as his beams decline."—*Somerville.*

> " Thick in yon stream of light, a thousand ways,
>   Upward, and downward, thwarting, and convolv'd,
>   The quivering nations sport ; till, tempest-wing'd,
>   Fierce Winter sweeps them from the face of day.
>   Even so luxurious men, unheeding, pass
>   An idle summer life in fortune's shine,
>   A season's glitter !　Thus they flutter on
>   From toy to toy, from vanity to vice ;
>   Till, blown away by Death, Oblivion comes
>   Behind, and strikes them from the book of life."—*Thomson.*

> Riches like insects when concealed they lie *
> Wait but for wings and in their season fly.—*Pope.*

The snail, butterfly, spider, lizard, and the rest, addressed
in some places as reptiles and vermin, are in others apostro-
phised as insects.　Again, all are described as emanating
alike from putrefying vegetable matter in hot weather.

> " Swampy fens
> Where putrefaction into life ferments,
> And breathes destructive myriads.
>
> The hoary fen
> In putrid streams emits the living cloud
> Of pestilence." ·

Especially, in a dozen poets at least, from the Nile mud :—

> " When thus, the Nile, diffus'd his wat'ry train
>   In streams of plenty o'er the fruitful plain,
>   Unshapen forms, the refuse of the flood
>   Issu'd imperfect from the teeming mud,
>   But the great source and parent of the day,
>   Fashion'd the creature, and inform'd the clay."—*Groome.*

This prolific diluvion, produces indeed, a very large variety
of zoological species, from the crocodile to the mosquito.
About the hippopotamus I will not be certain.　But it is

remarkable how all the poets agree about Nile mud produc-
ing misshapen and monstrous forms.   Pope is by no means
alone in his "half-formed insects on the bank of Nile."

But, poetically speaking, "insects"—"the mixing
myriads of the setting beam"—differ from reptiles and
vermin in this, that they are pitiably ephemeral.   They are
"a daily race"—

> "Swarming in the noontide bower,
> Rise into being and exist an hour."—*Darwin.*

They live such a short time that the poets generously excuse
them, as "beings of a summer's day."   This condescension,
I think, is hardly called for.   Thomson, for instance, is
good enough to say that "the ceaseless hum" in the woods
at noon is "not undelightful."   He will not say outright
that it is delightful, but to show what a large-hearted poet
he is, how impressionable to the sounds of wild nature, vows
half apologetically that, speaking for himself—he will not
be answerable for other tastes and does not wish to force
his own upon his readers—he does really, upon his honour,
and all joking apart, find something almost agreeable in the
humming of bees in summer woodlands!   What a generous
admission!   How such a confession draws the hearts of all
lovers of nature to the poet!   But let us hear him again :—

> "Nor shall the muse disdain
> To let the little noisy summer race
> Live in her lay, and flutter through her song."

What a beautiful condescension have we here!   How exquisitely tender!   He, Thomson—do not laugh at him,
ladies and gentlemen ; it is his gentle nature makes him do
it—will positively, and of his own accord, mention in his
beautiful poems such vermin as grasshoppers and butterflies.
"The muse," forsooth !

Or take Cowper's lines in the "Task : "—

> " I would not enter on my list of friends
> (Though graced with polish'd manners and fine sense,
> Yet wanting sensibility) the man
> Who needlessly sets foot upon a worm.
> An inadvertent step may crush the snail
> That crawls at evening in the public path ;
> But he that has humanity, forewarn'd,
> Will tread aside and let the reptile live.
> The creeping vermin, loathsome to the sight,
> And charged perhaps with venom, that intrudes,
> A visitor unwelcome, into scenes
> Sacred to neatness and repose, the alcove,
> The chamber, or refectory, *may die :*
> A necessary act incurs no blame.
> Not so, when held within their proper bounds,
> And guiltless of offence, they range the air,
> Or take their pastime in the spacious field :
> There they are privileged, and he that hunts
> Or harms them there is guilty of a wrong,
> Disturbs the economy of Nature's realm,
> Who, when she formed, designed them an abode."

That a man living in the country should be so indifferently informed, and though of a poetical turn of mind so unsympathetic, is almost unintelligible. What manner of thing does Cowper mean by "creeping vermin, loathsome to the sight, and charged perhaps with venom"? The poet means the toad, which "the vulgar" believe to be poisonous. But every gardener knows that it is a most useful little creature, and it is not vermin in any sense of that word. Anyhow, I cannot admire the "sensibility" of the poet who confesses that he approves of killing toads because they come into "the alcove," nor the "humanity" that draws a line between the needful and the needless treading on worms. To kill a toad simply because it comes into a summer-house is stupid cruelty.

CHAPTER II.

CROCODILES, TURTLES, AND LIZARDS.

" THE crocodile, the dragon of the waters, In iron panoply fell as the plague, And merciless as famine, " is obviously a creature that no poet can be expected to admire.   And it would perhaps be stretching sentiment too far to expect them to do so.   It is not a lovable beast.   I have seen them, huge ones, lying on a mud-bank, "like a forest-tree, basking in the sun," as Mary Howitt says, or crawling through reeds, and there was something in the demeanour of the thing that always made me long to kill it.   It lay flat, with a sluggish affectation of humility that exasperated me, and be-stirred itself with an air of helplessness that was positively monstrous.

A remarkable passage in Montgomery's "Greenland" shows us a broad river "swarming with alligator * shoals" and rolling "clouds of blood."   Thomson has a " Behemoth " that, "in plaited mail, rears his head "—"glanced from his side, the darted steel in idle shivers flies "—and that " crops upon the hills his varied fare."   That the former knew what he was writing about is as certain as that the latter did not, yet each is a conundrum.   When very young, crocodiles do certainly go "in shoals."

I have myself, in the Ganges' overflow, within eyesight from my house in Allahabad, seen them so thick that their

* " Alligator," it should be noted, is in poetry an aggravated crocodile. It is what the scritch-owl is, among birds, to the owl. —P. R.

opened jaws—they lie with their chins upon the bank and mouths open, perhaps for the same reason that you see vultures sitting facing the breeze with wings widespread—looked like a fringe of hurdle stakes, or *chevaux de frise.* At Dholpur, near Agra, I have also seen them, full grown, in a large company.   But these were retainers of the temples. As Bubastis had its cats, and Thebes her ibis, and Memphis her horned divinities, so Dendera had its collared crocodiles :

> " Clouds of incense woo thy smile,
> Scaly Monarch of the Nile."

The priests put necklaces about their necks, and earrings in their ears, and the people worshipped them, these grim obscurities that crept from out the depths of the river to take their place under Isis' throne.   But why was Montgomery's " broad river " red with blood?   The alligator takes his victim down, and there is no sign of the tragedy. A few pink bubbles perhaps—but that is all.

For Thomson's Behemoth I have no respect.   He had very vague or very confused ideas about the crocodile—which is " Leviathan " and not Behemoth—the rhinoceros, and the hippopotamus.   But, whatever it was, his picture is absurd.   For crocodiles do not eat grass.   The rhinoceros does not live in the water.   The hippopotamus is not mailed.

Keats, as usual, is true to Nature.   His " encasèd crocodile " is sufficient, and when he adds a reverence—

> " Son of the old Moon-mountains African
> Stream of the Pyramid and crocodile,"

the conjunction is worthy of the brute "in adamantine scales, That fears no discipline of human hands."

How Job militant exults in Leviathan :—

> " Dissect the greatnesse of so vast a Creature,
> By view of severall parts summe up his feature :
> Whe shields his scales an plac't, which neither art
> Knowes how to sunder, nor yet force can part.

> Ne belching rucks forth flames, his moving eye
> Shines like the glory of the morning skie ;
> His craggy sinewes are like wreathes of brasse
> And from his mouth quick flames of fier passe
> As from an Oven ; the temper of his heart
> Is like a Nether-Milstone, which no Dart
> Can pierce, secured from the threatning Speare ;
> Afraid of none, he strikes the world with feare.
> The Bow-moms brawny arms send shafts in vaine,
> They fall like stubble, or bound backe againe :
> Stones are his pillow, and the Mud his Downe,
> In earth none greater is, nor equall none,
> Compar'd with him, all things he doth deride
> And well may challenge to be King of Pride."—*Quarles.*

But why does Thomson describe the great beast as "cased in green scales," or Shelley imagine the species to have been exterminated by the Deluge ?—

> "The jagged alligator, and the might
>   Of earth-convulsing behemoth, which once
>   Were monarch beasts, and on the slimy shores
>   And weed-overgrown continents of earth
>   Increased and multiplied like summer worms
>   On an abandoned corpse, till the blue globe
>   Wrapt deluge round it like a cloke, and they
>   Yelled, gasped, and were abolished."

That this reptile, "who can falsely weep," as Heber says,[1] is a hypocrite, who needs telling? The poets are much attracted by this fancy. "With a feigned grief the tomb relents, And like a crocodile its prey laments," says Congreve. In Savage we find it "weeping cruel tears" over its "bleeding prey." And in Thomson it is "the smooth crocodile Destruction." Coleridge gives Hypocrisy a "crocodile's eye ;" and Shelley in the "Masque of Anarchy," sees her ride by on its back.

Spenser draws from the saurian's "swike" the admirable moral that it is as well to mind your own affairs while charitably bent on minding those of others :—

---

[1] "And the beeste who can falsely weepe
    Crocodilus was here goode shepe."

> "As when a weary traveller, that strays
>   By muddy shore of broad seven-mouthed Nile,
>   Unweeting of the perilous, wand'ring ways,
>   Doth meet a cruel, crafty crocodile,
>   Which, in false grief hiding his harmful guile,
>   Doth weep full sore, and sheddeth tender tears ;
>   The foolish man, that pities all this while
>   His mournful plight, is swallowed up unawares.
>   Forgetful of his own, that minds another's cares."

Southey puts into doggerel the legend of there being a king of the crocodiles. A woman has her child eaten up by one of the great river-lizards, and determines to complain to their ruler :—

> " The King of the crocodiles never does wrong,
>   He has no tail so stiff and strong,
>   He has no tail to strike and slay,
>   But he has ears to hear what I say."

So she goes, against much advice (like the youth in Excelsior) and eventually finds herself in the presence of the master monster :—

> " The King of the crocodiles there was seen,
>   He sate on the eggs of the crocodile queen,
>   And all around, a numerous rout,
>   The young prince-crocodiles crawled about."

And then she appeals to his Majesty in the words quoted above, and having listened to her—

> " You have said well, the King replies,
>   And fixed on her his little eyes ;
>   Good woman, yes, you have said right,
>   But you have not described me quite.
>   I have no tail to strike and slay,
>   And I have ears to hear what you say,
>   I have teeth moreover, as you may see,
>   And I will make a meal of thee,"

which he promptly does.

There is one more point about the "natural history" of the great reptile that is worth noting, namely, the incongruous friendship of the gruesome beast with the pretty little "zic-zac" plover.   Moore has the following :—

> "The puny bird that dares with teasing hum
>   Within the crocodile's stretched jaws to come."

Young has—

> "Like the bird upon the banks of Nile,
>   That picks the teeth of the dire crocodile."

and Spenser this :—

> "Beside the fruitful shore of muddy Nile,
>   Upon a sunny bank outstretched lay
>   In monstrous length a mighty crocodile,
>   That, crammed with guiltless blood, greedy prey
>   Of wretched people travelling that way,
>   Thought all things less than his disdainful pride,
>   Soon came a little bird called tidula,
>   The least of thousands which on earth abide,
>   That forced this hideous beast to open wide
>   The grisly gates of his devouring hell,
>   And let him feed as nature doth provide
>   Upon his jaws that with black venom * swell.
>   Why then should greatest things the least disdain,
>   That so small so mighty can constrain?"

It is probably only once in Keats, that crocodiles are an incident of a pretty scene :—

> "Onward the tiger and the leopard pants,
>   With Asian elephants :
>   Onward these myriads—with song and dance,
>   With zebras striped and sleek Arabians prance,
>   Web-footed alligators, crocodiles,
>   Bearing upon their scaly backs, in files,
>   Plump infant laughers mimicking the coil
>   Of seamen, and stout galley-rowers' toil :
>   With toying oars and silken sails they glide,
>   Nor care for wind and tide."

The "low-roofed" Tortoise meets with but scanty compliments from poets.   There is an unexpected sympathy,

* Query, "vermin."

ever, with it in the poetical conceit of the young turtle
born on the dry land longing for the water. "The sad
tortoise for the sea doth moan," says Marvell. Another has,
"Sighing for the deeps like the turtle." Byron thus notices
the contrast :—

> " Here the young turtle, crawling from his shell,
> Steals to the deep wherein his parents dwell.
> Chipped by the beam, a nursling of the day,
> But hatched for ocean by the fostering ray."

Montgomery also, after telling how the parent reptile "steals
out at eve," explores the shore "with trembling heart," and
lays her eggs in the loose warm sand, goes on to describe
the escape of the happy youngsters that "by instinct seek
the sea : "—

> " Nature herself with her own gentle hand
> Drops them one by one into the flood,
> And laughs to behold their antic joy
> When launched in th' element."

This is all pleasant reading, for it shows a tender appreciation of the creature's natural life, "where, in fleshy mail the
tortoise climbs the rocks."

More than one poet makes the curious error of thinking
that turtles shed their shells, as, for instance, Garth, who
has, "There the tortoise hung her coat of mail."

As the creature that gives to civic feasts what Southey
calls "the fat of verdant hue," so dear to the aldermanic
palate—

> "Gorgonius sits, abdominous and wan,
> Like a fat squab upon a Chinese fan.
> He snuffs far off the anticipated joy ;
> Turtle and venison all his thoughts employ."—*Cowper.*

—it cannot escape favourable recognition. Says Byron
"The turtle-shell which bore, A banquet in the flesh it
cover'd o'er ;" and Churchill—

B

> " The turtle of a great and glorious size,
>   Worth its own weight in gold, a mighty prize
>   For which a man of taste all risks would run ;
>   Itself a feast, and ev'ry dish in one."

Though a creature to laugh over when we see it creeping stealthily about on tip-toe, as if it were abroad for the purpose of picking pockets, it has a very notable place in myth, and was almost universally reverenced. The East believes that the world rests upon a tortoise, which rests upon nothing—and what a grand old testacean it is, this Vedic turtle, standing simply on its own dignity, and yet upholding upon its Atlantean carapace all the burdens of the round world, and of them that dwell therein! Here is a subject for Walt Whitman himself, the self-sufficient, democratic, thewy-and-sinewy, double-sexed, bully-for-you, old tortoise. More power to your shell, sir! We creeping things take off our hats to you, testudinous ancient. And how splendidly the deliberate thing looms out of Hindoo myth as the hereditary foe of the mystical elephant, the Darkness.

The Red Indian to this day says that in the beginning of things there was nothing but a tortoise. It brooded upon space : covered Chaos as with a lid. But after a while it woke up : its solitary existence was irksome to it, and it sank splendidly into the abysmal depths ; and lo ! when it re-emerged, there was the terrestrial globe upon its back ! For something to do, it had fished up our earth from the depths in the protoplasmic fluids, and, rather than be idle, it still keeps on holding it up. But some day it will sink again, and then will come the End—with Ragnarok and Armageddon.

In Greek and Roman fancies, the tortoise hardly fares so well. It is the form to which a bright nymph, who had jested at the nuptials of Zeus and Heré, was turned into by Mercury ; and ridicule falls upon the greatest of the Greeks when a tortoise falls upon his head. Yet they, too, knew of the tradition of the world-supporting thing, and did reverence to it. And so from East to West, from antiquity to to-day, the creature, vast, ponderous, inert, has commanded and commands the homage of men.

Nor in poetry do the historic traditions of the creature altogether fail of notice. How Æschylus was killed every one knows, but in Spenser we find Thomalin (moralising on good and bad shepherds) localising the event in England, and making the victim of the eagle's mistake "a proud and ambitious pastour," by name Algrind, who lived in his own neighbourhood :—

> " One day he sate upon a hill
>   As now thou wouldst by me ;
> But I am taught by Algrind's ill
>   To love the low degree :
> For sitting so with barèd scalp,
>   An eagle soarèd high,
> That, weening his white head was chalk,
>   A shell-fish down let fly.
> She weened the shell-fish to have broke,
>   But therewith bruised his brain ;
> So now astonied with the stroke,
>   He lies in grievous pain."

Henceforth Thomalin refuses ever to go up to the top of a hill, lest an eagle with a tortoise should happen to be overhead. An excellent simile, drawn from a most unpromising source, is Moore's—

> " Raised the hopes of men as eaglets fly
>   With tortoise aloft into the sky,
> To dash them down again more shatteringly ! "

Of the connection of the tortoise-shell with the first lyre Shelley, among others, takes notable cognisance in his "Hymn to Mercury." The poet sees the child playing about outside the cave and chancing upon a tortoise :—

> " The beast before the portal at his leisure
>   The flowery herbage was depasturing ;
> Moving his feet in a deliberate measure
>   Over the turf."

He cries, "A treasure!" and, laughing, addresses the animal :—

> " A useful godsend are you to me now—
>   King of the dance, companion of the feast,
>   Lovely in all your nature!   Welcome, you
>   Excellent plaything!   Where, sweet mountain beast,
>   Got you that speckled shell?   Thus much I know,
>   You must come home with me, and be my guest;
>
>   You will give joy to me, and I will do
>   All that is in my power to honour you.
>   Better to be at home than out of door;
>   So come with me, and though it has been said
>   That you alive defend from magic power,
>   I know you will sing sweetly when you're dead.
>   Thus having spoken, the quaint infant bore—
>   Lifting it from the grass on which it fed,
>   And grasping it in his delighted hold—
>   His treasured prize into the cavern old."

Arrived there, he "featly" scoops the shell out, drills holes in it, fastens reeds into them, spreads leather across, fixes the cubits in, "Fitting the bridge to both, and stretched o'er all, Symphonious cords of sheep-gut rhythmical."   When he had finished "the lovely instrument," he tried it, and—

> " There went
>   Up from beneath his hand a tumult sweet
>   Of mighty sounds."

Montgomery gives the earliest Christian origin of the lyre :—

> " A shell of tortoise, exquisitely wrought
>   With hieroglyphics of embodied thought;
>   Jubal himself enchased the polished frame,
>   And Javan won it in the strife for fame;
>   Among the sons of music, when their sire
>   To his victorious skill adjudged the lyre."

The line, "That you alive defend from magic power," is worth a note.   The blood of the tortoise was considered by the ancients an antidote to subtle venom.   Protected itself by its shield, it became a protector.   The Romans bathed new-born infants in the shells. Its appearance unexpectedly was a very auspicious omen, as being traditionally opposed to the diabolical and mischievous. In England, as a thing of magic, it was part of the stock-in-trade of the alchemist, astrologer, and quack : so, in "Romeo and Juliet"—"In his needy shop a tortoise hung." As affording a shield, it has honourable associations referred to by several poets. Thus Rogers' lines, "The warrior's lance Rings on the tortoise with wild dissonance," reminds the reader of the device by which they kept from old Chronos the intelligence of the birth of Zeus, and of the challenge "on the ringing tortoise" of the Knight of Thrace. That the elephant and tortoise should be at perpetual feud, each considering himself the lord of the lake, is one of the funniest myths I know.

"Both animals (sun and moon) frequent the banks of the same lake, and have conceived a mortal dislike one for the other, continuing in their brutal forms the quarrel which existed between them when they were not only two men but two brothers. As the elephant and the tortoise both frequent the shores of the same lake, they mutually annoy each other, renewing and maintaining in mythical zoology the strife which subsists between the two mythical brothers who fight each other for the kingdom of heaven, either in the form of twilights, or of equinoxes, or of sun and moon." Yet they meet after all in the "Rape of the Lock" in more friendly rivalry :—

> " The tortoise here and elephant unite,
> Transformed to combs, the speckled and the white."

In poetical metaphor, as in mythology, the tortoise represents the lazy and slow :—

> " I would not be a tortoise in his screen
> Of stubborn shell, which waves and weather wear not.
> 'Tis better, on the whole, to have felt and seen
> That which humanity may bear, or bear not."—*Byron.*

But, as Bacon says, " slowness is not sloth." It would be harder work to walk a mile behind a snail than to run one after a hare. But " the tortoise-foot " is an established phrase with the poets when they wish to imply " sluggish." " The lazy tortoise," says Faber.

Do you remember, in Crabbe, the girl who gets weary of the sleek, overcautious vicar ?

> " The wondering girl, no prude, but something nice,
> At length was chill'd by his unmelting ice ;
> She found her tortoise held such sluggish pace
> That she must turn and meet him in the chase ;
> This not approving, she withdrew till one
> Came who appear'd with livelier hope to run."

In heraldry we find the Medici with the impress of a tortoise " under full sail," with the motto *festina lente.* The conceit, though not original, is excellent, and may be classed with the dolphin (emblem of celerity) and anchor of Vespasian, the fish and chameleon of Pope Paul, or the crab and butterfly of the Emperor Augustus.

A wonderful family is that of the lizards—the ancestors of the birds, and the sliding link between the snake and crocodile. Was there ever palimpsest or papyrus so fascinating, so engrossing, so important, or so accurately authentic, as that stone from Saxony on which the archæopteryx has left the complete record of itself stamped on the soft slab ? It is nothing less than its whole body. Could any chronicle be more simple, unequivocal, satisfying ? How ingenuously it appeals to our confidence. No room is left for disputing its facts or cavilling at its arguments. There it lies as flat as the pressure of some millions of tons of overlying rock could make a thing, a shadow in thickness. Yet that little skeleton speaks with a logic that is most masterful, commanding, and unanswerable. It is the thing itself, crying out to us from the dim past—a phantom from the Genesis. Its speech is silence, and yet august in dumbness, its voice is more than trumpets; the walled-in cities of old superstition, the sacred citadels of ignorance, topple into ruins before it.

It is a word straight from the Demiurge himself, whispered to us through the rock galleries that stretch back from Now to Then—a single word spoken from the " In the beginning " —a worshipful thing. I never go to the British Museum without passing the model of this archæopteryx, the first of the birds. The lizard-fowl is a perpetual reverence to me.

Yet again, contemplate the way in which these creatures gradually shorten their legs as skinks, lose them altogether as amphisbænas, hesitate for a while as blindworms, and then become actual snakes—ophidian, viperine, terrible !

They commence with the pretty agile little things of our English sandy heaths, that are the "beasts-of-prey" to the tiny fly-folk who range among the grass as among forests, and find their lakes in dewdrops, their pleasure-parks on plaintain leaves. To them succeed the amphibious lizards of the New World, who feed on small snakes, mice, and birds—larger creatures, a yard in length, splendidly painted with yellow on a black ground. So to the water-lizards of the Old World, the crocodile-like " monitors," which—so tradition used to fancy—whistled a note of warning to Leviathan when danger threatened, albeit it eats the crocodile's eggs, and the young ones too. In revenge, the old crocodiles eat it. Next is that wondrous family of the " short-tongued " lizards, the arboreal iguanas, contrasting notably in their fearsome appearance with the floral loveliness of their tropical woodland haunts. Here, too, is the basilisk—the "dragon" of the Middle Ages in miniature— heraldic, grotesquely heterodox ; and the sea-lizard, a dread· ful-looking thing that feeds upon the sea-weeds a mile from

the coast; and the flying lizards, beauteous beyond description, that slide through the air from tree to tree on their
wing-like parachutes; and, most curious perhaps of all, the
frilled lizard, which, if it were only the size of a camel,
might have frightened all the Seven Champions out of their
wits. And what can be said too enthusiastic for such
a thing as the Moloch, a mass of spines and prickles,
with forty horns on the tip of its nose, and ferociously
thorned to the tip of its tail? or the geckos, the familiar but
wondrous creatures that sleep all through a summer's day
upon the ceiling and never drop off, but if they are startled
drop their tails? or the chameleon, that has such a transparent skin that its emotions can be read through it?

Human beings have been known to " blush crimson " or
turn "deathly pale." The choleric man turns vicious red
when out of temper, the Asiatic green when terrified. But
the chameleon beats us all. It has no expression whatever
on its face, so it makes up for it on its body. You can tell
what it is thinking about by the colour of its body. How
the poets delighted in the creature !

> " As the chameleon, who is known
> To have no colours of his own,
> But borrows from his neighbour's hue
> His white or black, his green or blue,
> And struts as much in ready light,
> Which credit gives him upon sight,
> As if the rainbow were in tail
> Settled on him and his heirs male."—*Prior.*

Allan Ramsay adapts an old fable excellently in his poem
on the beast. One man swears it is blue. He saw it that
very morning, and so cannot be wrong. Another had seen
it that evening, only an hour ago, and he will stake his life
on it, it was green. From argument it comes to quarrelling,
and "frae words there had been cuff and kick," but a third
man happens to come along. He asks the reason for such
high words between neighbours, and they tell him. At

this he laughs immoderately, calls them both fools, and says the chameleon is black, and he knows it—why?— because he has got it in his pocket at that very moment. And he whips the creature out—

> " But to surprise them ane and a'
> The animal was white as snaw ! "

Shelley too has an exhortation admirably characteristic :—

> " Chameleons feed on light and air ;
> Poets' food is love and fame.
> If in this wide world of care
> Poets could but find the same
> With as little toil as they,
> Would they ever change their hue
> As the light chameleons do,
> Suiting it to every ray
> Twenty times a day?
>
> Poets are on this cold earth
> As chameleons might be,
> Hidden from their early birth
> In a cave beneath the sea.
> Where light is chameleons change ;
> Where love is not poets do.
> Fame is love disguised : if few
> Find either, never think it strange
> That poets range.
>
> Yet dare not stain with wealth or power
> A poet's free and heavenly mind.
> If bright chameleons should devour
> Any food but beams and wind,
> They would grow as earthly soon
> As their brother lizards are.
> Children of a summer star,
> Spirits from beyond the moon,
> Ah ! refuse the boon ! "

But in its natural aspect the poets knew little of it. They pretended to believe it "fed" upon air :—

> " Stretch'd at its ease, the beast I view'd,
> And saw it eat the air for food."—*Herrick.*

> "On that hope,
> I build my happiness, I live upon it,
> Like the chameleon, on its proper food,
> The unsubstantial air."—*Hurdis.*

> "Bards are not chameleons quite,
> And heavenly food is very light."—*Montgomery.*

> "No living thing, whate'er its food, feasts there,
> But the chameleon who can feast on air."—*Churchill.*

> "While I, condemned to thinnest fare,
> Like those I flatter live on air."—*Gay.*

"Cold" is Sir William Jones's epithet—borrowed, of course, from the general lizard idea, that these creatures are gelid—

> "E'en cold chameleons pant in thickets dun,
> And o'er the burning grit the unwinged locusts run."

It was thus that the salamander got a reputation for disregarding flames, or even putting them out, by the extreme "coldness" of its body. It is therefore in metaphor that this strange lizard is most conspicuous. The gay gallant in Moore—

> "Pranked in gay vest, to which the flame
> Of every lamp he passed, or blue,
> Or green, or crimson, lent its hue;
> As though a live chameleon's skin
> He had despoiled to robe him in."

the turncoat politician in Churchill—

> "A creature of the right cameleon hue,
> Wears any colours, yellow or true blue."

In Cowley Fancy, in Savage Fortune, has the same steeds —

> "Wild dame with much lascivious pride
> By twin chameleons drawn does gaily ride."

Advisers are chameleons in Dryden—

> "To change the dye, with every distant view."

Love in Shakespeare; courtiers in Gay; Italians in Rogers; the sea in Campbell :—

> " Mighty sea! chameleon-like thou changest."

lovers in Shelley :—

> " As a lover or a chameleon
>   Grows like what it lives upon ; "

cowards in Byron—

> " Took, like chameleons, some slight tinge of fear ; "

and women—"your true chameleons"—in Pope, "who change colour according to the humour and disposition of the men who approach them;" the weak-minded man in Pryor :—

> " As the chameleon who is known
>   To have no colour but his own,
>   But borrows from his neighbour's hue,
>   His white or black, his green or blue ;
>   And struts as much in ready light
>   Which credit gives him upon sight."—*Prior.*

> " Ah ! simple youth ! how oft will he
>   Of thy chang'd faith complain !
>   And his own fortunes find to be—
>   So airy and so vain,
>   Of so cameleon-like an hue,
>   That still their colour changes with it too ! "—*Cowley.*

> " Through a thousand shades
>   His spirit flits, chameleon-like, and mocks
>   The eye of the observer."—*Rogers.*

That they thrive on a diet of air is a point that is obviously suggestive :—

> " *Speed.* Why muse you, sir? 'Tis dinner-time.
> *Valentine.* I have dined.
> *Speed.* Ay, but hearken, sir ; though the chameleon Love can feed on the air, I am one that am nourished by my victuals, and would fain have meat."

And so in Hamlet :—

"*King.* How fares our cousin?
Hamlet. Excellent, i' faith—of the chameleon's diet: I eat the air, promise-crammed."

Once upon a time, says a Hindoo legend, there was a certain king who had before him a case in which two Brahmans disputed the possession of a cow and calf, and he was so dilatory in judgment that the pious litigants appealed to Heaven, the result, as far as the king was concerned, being that he was turned into a chameleon, which never seems to know its own mind for an hour together.

Poets see lizards in two aspects—either as things of happiest, brightest sunshine, or of ominous and sepulchral gloom. Those, as Byron, Shelley, Rogers, Montgomery, or Faber, who had seen and therefore admired these pretty, elegant, harmless creatures, speak of them with kindly admiration. They hear "the quick-eyed lizard rustling through the grass," or note "the shrill chirp of the green lizard's love," see "the lively lizard playing in the chinks," and watch it basking in the grooves of the fallen pillar—

"With sensual enjoyment of the heat,
　And with a little pulse that would outstep
　The notes of nightingales for speed."

On the other hand, they are creatures of ruins and dismal abodes—

"Bit by bit the ruin crumbles,
　Not a lizard there abiding;
And the callow raven tumbles
　From the loophole of her hiding."

"The painted lizard and the bird of prey" are associates in Dryden (borrowing from Virgil); in Cunninghame we have "the lizard and the lazy, lurking bat, Inhabiting the painted room." These are poets of the non-natural-history category, and supreme amongst them are the Eliza Cooks of verse. As, for instance, " Bat and lizard had allied, With mole and owlet by their side;" or "The dark retreat of lizard, frog, and speckled snake;" or "The gloomy owl and speckled lizard."

Shelley is especially fond of the lizard simile, and having in Greece and Italy had these beautiful creatures constantly under his eye, uses them in his verse with exquisite felicity, although at times with a "large license." Notable is the fidelity of his attaching the idea of "light" to the sudden-flashing thing. It is "sparkling" "glittering"—"and the green lizard and the golden snake, Like unimprisoned flames out of their trance awake."

Now, whimsical though it may seem, I should like to draw passing attention to the curious community of epithet which all these creatures enjoy.

We have "speckled" lizard with the owl, and the lizard with the "speckled snake," and the latter with the owl. There is the "painted" lizard and the owl, also "the painted snake and the owl." Then we read of "gay lizards glittering," and "serpents glittering, with gay hues adorned." The "green gilded snake" glides on the tomb, and the "green lizard and gilded newt" do the same on a ruin. A third has "a green gilded lizard." I could go on to tedious length, but my object in this brief paragraph is only to suggest that poets are immoral in lumping diverse creatures together in order to convey a particular impression. The mid-day lizard and the nocturnal owl are of course as absurd in association as the land lizard and the water newt.

Allan Ramsay has a poem on "Twa Lizards," which is zoologically interesting, though the moral is dull enough— after Spenser. Of two lizards basking on a bank, one regrets their mean estate, and cites the existence of crocodiles on the Nile, which are worshipped in "pagods,"

as an example of what his ambition aspires to.　Those were
lizards worth calling such—or he would even like to be a
deer, with fine horns.　Of course a deer is run down before
his eyes, and the lizard repents his aspirings after an exalted
station.　"It is better," says Herbert, "to be the head of a
lizard than the tail of a lion."

As one of the heralds of Spring, the "lizard of St. Agnes"
is a popular favourite in Southern Europe.　In Italy it is
also called "guarda uomo," man's protector.　In Sicily it
stands in equal favour, and San Giuvanni, as it is called,
must not be killed, "because it is in the presence of the
Lord in heaven, and lights the little lamps before the Lord."
But if by accident you should do one to death, you must
touch the still quivering limbs and say :—

> "Not I, not I, did murder you,
> Little dog of holy St. Matthew."

The children carry them about as pets in their bosoms,
and when they let them go, ask them to intercede before
"the Lord" for them.　The common green lizard is
especially protected, as superstition invests it with power
against evil talismans and enchantments, and against
venomous snakes.　Thus the crest of the Mantuan princes
was a lizard in a tuft of camomile, Pliny saying that these
creatures, when they have fought with serpents and been
bitten, cure themselves with that herb.

It is unlucky, so English folk-lore avers, for a wedding-
party to see a lizard ; and that the creature has a painful
sting—says Suffolk, " as smart as lizards' stings ; " and again,
Queen Margaret, "lizards' dreadful stings "—is still an
article of superstitious belief among the more ignorant.

The Arabs eat the legless desert lizards, calling them
"sand-fishes," so as not to seem to transgress commandment;
for the creatures are unclean, one species indeed being
specially anathema, for Mahomet has ordered his people

to stone them wherever found, as they hang their necks in mockery of the Moslem's attitude of prayer. To this day, therefore, the faithful persecute them rigorously as scoffing unbelievers.

" Fadladeen, besides the spiritual comfort he derived from a pilgrimage to the tomb of the saints from whom the valley is named, had opportunities of gratifying, in a small way, his taste for victims, by putting to death some hundreds of those unfortunate little lizards, which all pious Mussulmans make it a point to kill ;—taking for granted that the manner in which the creature hangs its head is meant as a mimicry of the attitude in which the faithful say their prayers ! "

None the less they were considered, and indeed are still, very important in Eastern medicine, the traditional coldness of the creatures recommending them to the Pharmacopœia of Fancy as being supposed to be beneficial in all ailments arising from excessive heat, although so diverse as burns, sunstroke, sand-blindness, and scarlet fever.

Newts, the pretty " eft " of our ponds, receive the most infamous treatment from poets. Wordsworth calls them " offensive," Eliza Cook miasmatic :—

> " Mist and chill are over the hill,
>   The crops on the upland are green and stark,
>   *Newts are about* and the rain puts out
>   The tender light of the glow-worm's spark."

It delights me to quote this fustian. Imagine crops being " stark ! "—and then that line, " *newts are about !* "

But Spenser calls them " fearefull eftes," Garth " hateful," and Shelley " poisonous." Shakespeare's witches mix the eyes of newts with the toes of frogs in their dreadful broth.

> " Onely these marishes and myrie bogs,
>   In which the fearefull ewftes do build their bowres,
>   Yeeld me an hostry mongst the croking frogs."—*Spenser.*

> " And from a stone beside a poisonous eft
>   Peeps idly."—*Shelley.*

> " Down to those cells obscener reptiles creep,
>   Where hateful newts and painted·lizards sleep,
>   And shiv'ring snakes the summer solstice wait."—*Garth.*

And here out of sheer malice, I will quote Wordsworth. whose errors of natural history and lack of sympathy with nature are as deplorable as inexplicable in a rural poet :—

> " How disappeared he? ask the newt and toad,
>   Inheritors of his abode ;
>   The otter crouching undisturbed,
>   In her dank cleft ; but be thou curbed,
>   O froward fancy ! 'mid a scene
>   Of aspect winning and serene ;
>   For those offensive creatures shun
>   The inquisition of the sun ! "

Why is the newt, a charming ornament of the waters it frequents, "offensive ;" and why does it shun the sunlight any more than trout or salmon do?

Again :—

> " The tapers shall be quenched, the belfries mute,
>   And 'mid their choirs unroofed by selfish rage,
>   The warbling wren shall find a leafy cage
>   The gadding bramble hang her purple fruit ;
>   And the green lizard and the gilded newt
>   Lead unmolested lives, and die of age."

Why "gilded" newt? The scarlet fringe of its tail is sufficiently conspicuous, and worth noting, but why gilded ? Why not electro-plated ?

CHAPTER III.

SNAKES IN NATURE.

IN all the range of poetry there is no object of nature, outside humanity, which has engaged fancy more constantly or in so many diverse moods as the serpent.   It was invested in Holy Writ with a most portentous individuality; has been reverenced at one time or another with divine honours by almost every race upon the earth, and coils inextricably round the legends of nearly every language.   It was endowed in classical literature with all conceivable attributes, malignant and benign, and honoured through successive ages with such persistent superstition as to constitute it almost the central figure of folk-myth.   Is it then to be wondered at that the poetic mind should be attracted by a creature which has fascinated mankind from the earliest times, and which still maintains its rank as the chiefest of Nature's parables?

The serpent, however, has a prodigious literature of its own, and into this I have here no intention to make rash expedition.   My concern is specifically with "Snakes in Poetry," and even when thus restricted, the subject is sufficiently large and many-sided, "rolling in orbs immense its length of coils," to make me prefer to take it in three sections.   The first of these is the reptile "in nature."

Now, very few of our British poets knew personally anything at all of the snake in nature, and this absence of information compelled them to go for the facts they wanted to Holy Writ, classical myths, or popular superstitions.   More-

C

over, in the matter of natural history they follow each other with remarkable fidelity, and nearly all the errors of later poets are to be traced, by the actual language used, back to the elder.   The serpent, therefore, continues to be "deceitful," and the adder "deaf," and snakes generally are said to be "slimy," to leap upon their victims, to wound with their tongues, or sting with their tails.   Like them there is nothing for fatal and determined malignity: it is death to enrage them.

Yet, how very different are the facts.   Even in my own casual acquaintance with venomous species I have learnt—indeed, the first experience was enough to teach me—what poor, helpless, timid, destructible creatures they are.

When living on the shore of the Great Salt Lake in Utah, I used to go hunting for rattlesnakes with a forked stick. The suicidal creatures give you but small trouble in "hunting," for as soon as you come near them they spring their alarum.   "Here I am," the reptile calls out.   So, guided by the sound, you discover under a tuft of sage-brush the object of your quest, and in spite of

<blockquote>"The rattling terrors of the vengeful snake,"</blockquote>

you proceed to fix it with the fork, and then with the heel of the boot or a stone to kill it, cutting off its rattles as a trophy.   The diversion at first even suggested cruelty to me: for the snake had no chance whatever, and made no effort worth calling such to escape.   But this foreshore was the favourite summer bathing-resort of the residents of Salt Lake City, and as the children wandered about in all directions picking flowers, the danger to life was so considerable (as had been disastrously proved), that the killing of rattlesnakes became as much a public service as the destruction of other venomous species—undertaken, for instance, by the Government of India—is considered elsewhere.   But when I hear any one speak of the terrors of the rattlesnake, I know they have never gone hunting them.

To give one more personal reminiscence. I was at one time a professor at the Agra College, and, sitting in my chair one hot-weather morning, was holding forth with something more than my usual earnestness to the small class before me. The subject was some problem in metaphysics, the students were undergraduates reading for honours, and examinations were close at hand. It was necessary we should all be very diligent. My annoyance was very considerable, therefore, when I saw that the half-dozen young fellows—some of them were Brahmans—persisted in looking over my shoulder instead of at me. At last I looked round to see what it was that distracted their attention, and found, to my surprise, that a passing cobra, attracted probably by the droning of my voice, had come into the lecture-room, and was enjoying itself in a corner. These creatures, as every one knows, are peculiarly susceptible to anything like cadence in sound, and it is very likely that the monotonous rise and fall of a single voice had taken its fancy, just as modulation of notes, without any air, upon any musical instrument will do. At any rate there was the cobra, and as fine a speciman as I ever met with, with its hood inflated to the full, its spectacles brilliantly white, and the sunlight striking in through the doorway across its burnished body. There is something singularly imposing in the attitude of this snake when excited. My visitor in the present case had raised itself as high as it could, something less than a foot, and was swaying from side to side in accurate rhythm, as if in a trance ; and in the uplift of the head, the proud drawing-back of the neck, there was a positive majesty of bearing hardly conceivable in a poor worm some four feet long. Now, a Portuguese author, writing of India, says: "The sudden appearance of a cobra-da-capello in a room is considered to presage some future good or evil. It is the Divinity himself in this form, or at least his messenger, and the bringer of rewards or chastisement. Although it is exceedingly venomous, it is neither killed nor molested in the house which it enters, but respected, and even caressed." Far different was the fate of the cobra that intruded upon my lecture-room. For I got up, and, keeping my ruler in my hand, went towards the snake. Turning to my class, I said, "It is a very sacred animal, I know—but not in a lecture-room." And therewith, while the poor creature was still continuing its sing-song oscillation, I knocked it over with a smart tap on the head from my ebony sceptre. "Besides," I continued, "the Government has placed a reward of fourpence-halfpenny on its head," and I took up my white sun-umbrella, which was leaning against the wall, and, putting the point of it under the writhing thing, jerked it clear out through the doorway into the sunlight. And before I got back to my chair there was a rush of wings to the doorway, and the next instant a couple of kites were carrying the cobra away in halves into two separate parishes.

So it will be understood that I have not for the natural serpent any superstitious reverence. On the other hand, it may be objected that I am not a fit person to undertake the criticism of poets on a subject so full of suggestive fancies. In self-defence, therefore, I venture to say that I have written much upon this fascinating subject, and not altogether, I hope, without sympathy with the beautiful myths of antiquity and the engaging credulities of more modern ignorance ; and to escape, therefore, the charge of not recognising the aspects from which poets survey the reptile world, I will repeat here a paragraph from a paper which I once took the liberty of putting into Mr. Ruskin's mouth as a lecture on snakes.

"Without a horizon on any side of him, the speaker could hold high revel among a multitude of delightful phantasies, and make holiday with all the beasts of fable. Ranging from Greek to Saxon, and from Latin to Norman, Mr. Ruskin could traverse all the cloudlands of myth and the solid fields of history, lighting the way as he went with felicitous glimpses of a wise fancy, and bringing up in quaint disorder, and yet in order too, all the grotesque things that heraldry owns, and the old world in days past knew so much of; the wyvern, with its vicious cast of countenance, but inadequate stomach; the spiny and always rampant dragon-kind; the hydra, unhappy beast that must have suffered from such a multitudinous toothache; the crowned basilisk, king of the reptiles and chiefest of vermin; the gorgon, with snakes for hair; and the terrible echidna; the cockatrice, fell worm, whose first glance was petrifaction, and whose second death; the salamander, of such subtle sort that he digested flame; the chimæra, shapeless, yet deadly; the dread cerastes; the aspic, "pretty worm of Nilus," fatal as lightning and as swift; and the dypsas, whose portentous aspect sufficed to hold the path against an army of Rome's choicest legion. From astronomy, where Serpentarius, baleful constellation, glitters, and Draco refulgent rears his impossible head, the speaker could run through all the forms of dragon idealism, recalling to his audience, as he went on his way, beset with "unspeakable" monsters, the poems of the Greek and of yet older mythologies, churning up the old waters with a Shesh of his own, and summoning into sight at the sound of his pipe all the music-loving reptiles of mythology, like one of the old Psylli or the Marmarids, or one of the Magi, sons of Chus, 'tame, at whose voices, spellbound, the dread cerastes lay.'

"The snakes of antiquity, it is true, have come down to us dignified, and made terrible by the honours and fears of past ages, when the Egyptians and the Greeks bound the aspic round the head of the idol as the most regal of tiaras, and crowned in fancy the adder and the asp; when nations tenanted their sacred groves with even more sacred serpents: entrusted to their care all that kings held most precious, and the gems which the jealous earth still held undug; deifying some of their worms, and giving the names of others to their gods. But the actual facts known to science of modern snakes, the deadlier sort of the ophidians, invest

them with terrors equal to any creature of fable, and with
the superstitious, might entitle them to divide honours with
the past objects of Ammonian worship and be the central
figures in the rites of Thermuthis or of Ops."

Very few specific varieties of the ophidian class find
notice in verse.   Apostrophising the Red Man—whom Mrs.
Hemans, by the way, calls "a snake"—Eliza Cook bids
him go and consort with "the whipsnake and the jaguar," a
task which is as cruelly severe as any ever set by the wicked
stepmothers of the fairy tales, for, to consort simultaneously
with both whipsnake and jaguar would necessitate the
Red Man's being in two places at once, seeing that those
creatures inhabit different continents.

The "rattlesnake"—the "Indian's rattlesnake" of Butler—
meets with frequent reference.   Byron has these mysterious
lines :—

> "Sprung from a race whose rising blood,
> When stirred beyond its calmer mood
> *And trodden hard upon*, is like
> The rattlesnake's in act to strike."

"Like a live arrow leapt the rattlesnake," says Montgomery; but Goldsmith's line, "The rattling terrors of the
vengeful snake," is perhaps the best, though Butler is
certainly more truthful to nature when he says—

> "One that idly rails and threats,
> And all the mischief that he meant
> Does, like a rattlesnake, prevent."

For this snake's alarum is, I think, from the personal experiences of the reptile already narrated, a merciful provision
for the security of man and beast, rather than any additional
circumstance of malignity in the reptile.   But for that warning sound I should myself have often come very near to
treading on them, and on one occasion actually touching
one with my hand ; but the smallest alarm makes the hidden
thing declare itself.   The noisy gift, in fact, is fatal to the
snake, and the salvation of everything else.

Following up her "whipsnake and jaguar" with another impossible association, Eliza Cook has "the boa and the vulture" consorting together for the Red Man's ruin.

The boa-constrictor is often alluded to, but mentioned by name only once again, unless I am mistaken, and then by Byron in Don Juan :—

> "Not that he was not sometimes rash or so,
> But never in his real and serious mood ;
> Then calm, concentrated, and still, and slow,
> He lay coil'd like the boa in the wood :
> With him it never was a word a blow ;
> His angry word once o'er, he shed no blood ;
> But in his silence there was much to rue,
> And his one blow left little work for two."

Asps, of course, have filled, ever since "the pretty worm of Nilus" hidden in fig-leaves was carried up by country clowns (momentous burden) into the palace of Cleopatra— herself Mark Antony's "serpent of old Nile"—a large space in serpent-lore. But they are not found often in English verse. In his "Camel-driver," Collins appropriately places it (the "parch'd adder" of Akenside) in Arabian deserts :—

> "At that dread hour the silent asp shall creep,
> If aught of rest I find, upon my sleep."

And perhaps in the next couplet he refers to the asp's natural companion in the sandy wilderness, the puff-adder :—

> "Or some swol'n serpent twist his scales around
> And wake to anguish with a burning wound."

That this little worm lets itself be eaten by cranes in order to feed at its ease upon the bird's entrails [1] is a curious fiction more than once alluded to in metaphor, and now and again the word "asp" occurs as a generic name for venomous snakes rather than of any specific viper. The "desert serpent" of Campbell that dwells in "desolation cold,"

---

[1] "As through the crane's trunk throat doth speed,
The asp who doth on his feeder feed."—*Lovelace.*

might be the asp or "horned cerastes dire " if the poet had not naturalised it in America.   But Campbell is at all times delightfully incorrect in his natural history.

King's "hornèd serpents " that Megæra wears, are perhaps the cerastes ; but Moore, misunderstanding the name of another species, makes an amusingly characteristic error as to its meaning.   He says :—

> "The smooth glass-snake, gliding o'er my way,
> Shows the dim moonlight through his scaly form "—

evidently thinking the reptile got its name from being transparent.   But it is as opaque as any other worm, and owes the prefix to its exceeding brittleness.   Moreover, the "glass-snakes " are hardly snakes at all, but only snake-like lizards.   Sir William Jones has an Indian "serpent dire," "of size minute, with necklace brown and freckled side," which is perhaps the *Daboia elegans*.   More uncertain is the "blue serpent" with which, in the satire of Rufinus, the Fury "girds her waist around," after binding her hair with *an adamant*.[1]

The water-snake, it might have been thought, would have been a very attractive image to poets, but such is not the case.   It is but seldom met with, and even on those infrequent occasions without any attempt to take advantage of so masterly a touch of nature.   Moore's fancy imagines many a "water-snake " slumbering in Lake Erie :—

> "Basking in the web of leaves
> Which the water-lily weaves."

But it is only in Shelley, the poet of the snake, that this serpent meets with competent recognition.   Thus :—

> "The snake,
> The pale snake that, with eager breath,
> Creeps here his noontide thirst to slake,

---

[1] The Fury, having performed these feats of the toilette, proceeds to Phlegethon, "whose *pitchy* waves are flakes of rolling flame !"

> Is beaming with many a mingled hue
> Shed from yon dome's eternal blue
> When he floats on that dark and lucid flood
> In the light of his own loveliness."

What "the deaf adder that stoppeth her ear: Which will not hearken to the voice of charmers, charming never so wisely," may have been in ancient Palestine, it is now scarcely possible to say ; but in the old version of Holy Writ the translators rendered the original word sometimes as "adder," sometimes as "cockatrice"—a fearful reptile, which in the days of King James was thoroughly believed in. But our poets, knowing that one of the English snakes is so called, transferred to it the epithet of "deaf"—regardless of the fact that the creature is really very quick of hearing. "Fierce" is another epithet sometimes coupled with it—"as the adder deaf and fierce;" "fierce as the adder and as deaf." Pope has "fierce as a startled adder," and our English reptile, though a timid thing, will, it is true, turn, but impotently, at bay if pursued and teased.

With regard to these two Biblical points, the "ferocity" and the deafness of the adder, the following passages from Wood's "Bible Animals" are worth quoting. Speaking of the general apathy of snakes, he says—

"The late Mr. Waterton, for example, would take up a rattlesnake in his bare hand without feeling the least uneasy as to the behaviour of his prisoner.

"He once took twenty-seven rattlesnakes out of a box, carried them into another room, put them into a large glass case, and afterwards replaced them in the box."

Coming then to the English adder, Mr. Wood gives the following personal reminiscence :—

"As a rule, a great amount of provocation is needed before a venomous serpent will use its teeth.

"One of my friends, when a boy, caught a viper, mistaking it for a common snake. He tied it round his neck, coiled it on his wrist by way of a bracelet, and so took it home, playing many similar tricks with it as he went. After arrival in the house, he produced the viper for the amusement of his brothers and sisters, and, after repeating his performances, tried to tie the snake in a double knot. This, however, was enough to provoke the most pacific of creatures, and in consequence he received a bite on his finger."

Respecting its deafness the same writer gives the following delightful quotation from a "Sermon for the eleventh day after Pentecost," by "Luis of Granada:"—

"Their fury is after the likeness of the serpent, as the asp which even stoppeth her ears—which heedeth not the voice of the charmers, even of the wizard which charmeth wisely.

"For they say commonly, the asp, while she is charmed, so that she poisoneth not men with her deadly venom, layest one of her ears to the ground and stoppeth the other by thereunto putting her tail, that so the strength of the poison which lurketh within may abide without."

To this the author adds the remark—

"It may be as well to remark, before passing to another of the serpents, that snakes have no external ears, and that therefore the notion of the serpent stopping its ears is zoologically a simple absurdity."

Viper is poetically, as also in popular language, a synonym of adder, and the name is given as a rule to all snakes that are of the smaller size but greater venom. To the larger, the name "serpent" is appropriated. To illustrate this from Thomson :—

> " Lo ! the green serpent, from his dark abode,
>   Which e'en Imagination fears to tread,
>   At noon forth issuing, gathers up his train
>   In orbs immense, then, darting out anew,
>   Seeks the refreshing fount ; by which diffused,
>   He throws his folds ; and while, with threatening tongue

> And deathful jaws erect, the monster curls
> His flaming crest, all other thirst, appalled,
> Or shivering flies, or checked at distance stands,
> Nor dares approach."

This is the serpent, the reptile of the largest size.   Then the passage continues thus :—

> " But still more direful he,
> The small close-lurking minister of fate,
> Whose high concocted venom through the veins
> A rapid lightning darts, arresting swift
> The vital current.   Formed to humble man,
> This child of vengeful Nature."

Here we have the two poetic genera in juxtaposition, the one terrific, bulky, crested, that awes all the wild things of the tropics by its furious aspect, and its acknowledged strength ; the other insignificant in size, but "lightning" in its deadliness.

Nor are these passages without interest as illustrating several very prevalent errors—prevalent, not only among poets—about this most wondrous order of creatures.   They may indeed be called the "normal" errors of the poets.

One of these is the idea that snakes—Titania's "spotted snakes with double tongues"—wound with their tongues. Shakspeare has both "tooth" and "sting," and was evidently in doubt on the point.   Thomson, we have seen, has "threatening tongue"—because, perhaps, Somerville (whom he had read assiduously and to much useful purpose) has the same expression :—

> " So when the unwary clown with hasty step
> Crushes the folded snake, her wounded parts,
> Grov'lling, she trails along, but her high crest
> Erect she bears, in all its speckled pride
> She swells, inflamed, and with her forky tongue
> Threatens destruction."

Scott compromises with his doubts :—

> " Thus, circled in his coil, the snake,
> When roving hunters beat the brake,

> Watches with red and glistening eye,
> Prepared, if heedless steps draw nigh,
> With *forked tongue* and venomed fang
> Instant to dart the deadly pang ;
> But if the intruders turn aside
> Away his coils unfolded glide,
> And through the deep savanna wind
> Some undisturbed retreat to find."

Many other poets are content to be equally ambiguous, while in some there is a suspicion of the further error that the snake "stings." With its tail ?   Marvell starts it, so far as I can gather, with—

> " Disarmèd of its teeth and sting ; "

and after him many follow, as Allan Ramsay in—

> " Th' envenomed tooth or forked sting ; "

or Eliza Cook in—

> "Crushing and stinging with venomed fold."

Philips continues the fiction with an admirable originality, hardly to be expected from the author of "Cider :"—

> " And as a snake, when first the rosy hours
> Shed vernal sweets o'er ev'ry vale and mead,
> Rolls tardy from his cell obscure and dank ;
> But, when by genial rays of summer sun
> Purg'd of his slough he nimbler threads the brake,
> *Whetting his sting*, his crested head he rears,
> Terrific from each eye retort he shoots
> Ensanguin'd rays—the distant swains admire
> His various neck and spires bedropp'd with gold."

The idea of "whetting his sting" is as delightful, but not so original as the rest, for in other poets we have the wild-boar and the rhinoceros whetting their tusks under very similar circumstances.   Moreover, there is the high prescription of Holy Writ : "They have sharpened their tongues like a serpent," says the Psalmist.   Southey makes thankful use of it in the lines—

> " Wily as the snake
> That sharps his venomed tooth in every brake."

So much for the errors as to what the Americans call "the business ends" of the snake.   Another class concerns itself with the creature's appearance.   "The vulgar" always call snakes "slimy," and poets do so too.   Thus Rogers, who ought to have known better, says :—

> " Everywhere from bush and brake,
>   The musky odour of the serpents came,
>   Their *slimy track* across the woodman's path
>   Bright in the moonshine."

The origin of the error that snakes are slimy, so far as modern poets are concerned, is perhaps Darwin, who more than once speaks of the "foamy folds" of serpents, and as he was a naturalist, his word of course went for much. Among other misconceptions as to the tribe may be noted Pitt's idea, that serpents feed on poison-plants—

> " So the fell snake rejects the fragrant flow'rs
>   But every poison of the field devours ; "

and the more common ones that these reptiles stand on end when angry, and that they are most active at noon and asleep by night.

Darwin has a remarkable fancy on the reciprocity of alarm :—

> " Stern stalks the lion ; on the rustling brinks
>   Hears the dread snake, and trembles as he drinks,
>   Quick darts the scaly monster o'er the plain,
>   Fold over fold his undulating train ;
>   And bending o'er the lake his crested brow,
>   Starts at the crocodile, that gapes below."

What power there is in Spenser's simple lines—

> " Like a snake whom wearie winter's teene
>   Hath worne to nought, now, feeling summer's might,
>   Casts off his ragged skin and freshly doth him dight."

All poets are attracted by this idea of rejuvenescence, and the casting of the slough is as regularly recurrent as

the deer's "hanging of his old head on the pale." Says
Somerville :—

> " Brisk as a snake in merry May
> That just had cast his slough away."

And Montgomery :—

> " The serpent flings his slough away
> And shines in Orient colours dight,
> A flexile ray of living light."

Whether or not these reptiles exercise a fascinating influence over other creatures is still an undecided point.
But antiquity held that they could charm with the eye; and
the bird spellbound by the snake has passed into an
accepted metaphor.   In verse its occurs abundantly :—

> " It was vain to hold the victim,
>   For he plunged to meet her call,
> Like the bird that shrieks and flutters
>   In the gazing serpent's thrall."—*Campbell.*

> " As the snake's magnetic glare
> Charms the flitting tribes of air,
> Till the dire enchantment draws
> Destined victims to his jaws."—*Montgomery.*

> " Like the bird whose pinions quake,
> But cannot fly the gazing snake."—*Byron.*

> " Thou'lt fly?   As easily may victims run
> The gaunt snake hath once fixed eyes upon ;
> As easily, when caught, the prey may be
> Plucked from his loving folds as thou from me."—*Moore.*

> " This cold and creeping kinsman who so long
> Kept his eye on me, as the snake upon
> The fluttering bird."—*Byron.*

Of their personal beauty the poets draw an almost
exaggerated picture.   Wondrous as the elegance and adorn-

ment of these creatures undoubtedly are, it is almost excessive to speak of their "volumes of scaly gold" and "thousand mingling colours," when referring to the actual reptile in nature. That Keats should make his Lamia transcendent in splendour, or Shelley his serpents of fancy such miracles of loveliness, is well within their licence; but when the real creature is under description, poetical rapture often goes beyond the subject. As in Montgomery :—

> " Terribly beautiful, the serpent lay
> Wreathed like a coronet of gold and jewels
> Fit for a tyrant brow ; anon he flew,
> Straight as an arrow shot from his own rings,
> And struck his victim shrieking, ere it went
> Down his strained throat, the open sepulchre."

Their eyes are not like "live rubies " nor "living emeralds,"

> " The light of such a joy as makes the stare
> Of hungry snakes like living emeralds glow,
> Shone in a hundred human eyes."—*Shelley.*

but, on the contrary, are most malignantly, venomously, dull, and quite incapable of "flinging out arrows of death."

That snakes leap at their victims is one of those popular errors which it seems impossible to destroy. For, as a rule, men and woman lose some of their presence of mind when confronted suddenly—and the snake is very sudden in its gestures—with one of these reptiles, and, if struck at, always declare that the creature "sprang " at them. But it is a fact that no snake can leave the ground; moreover, that the radius of their stroke is limited in a fixed relation to their length, a four-foot individual, for instance, being only able to wound at say a foot-and-a-half, and so on in proportion to the varying lengths. So that the snake

> " Who pours his length
> And hurls at once his venom and his strength "

is a poetical fiction, as, for the same reason, is Montgomery's brilliant reptile quoted above.

Scott has " Like adder darting from his coil," and Byron,
" As darts an angry asp."

Sometimes there is a harmless snake, as in Joanna
Baillie's "Devotional Song for a Negro Child," where
" stingless snakes entwisted lying," are mentioned among
the usual features of a tropical noontide—a very curious
effort of fancy—or, as in Waller's address to "A Fair Lady
playing with a Snake : "—

> " Thrice happy snake ! that in her sleeve
> May boldly creep ; we dare not give
> Our thoughts so unconfined a leave.
>
> Contented in that nest of snow
> He lies, as he his bliss did know,
> And to the wood no more would go ; "

and again, in Broome, in the poem "To a Lady," that commences, " It is a pleasing, direful sight, At once you charm
us and affright."   This more amiable aspect of the reptile
is, however, legitimately extended in the " Faëry Queen,"
where we find Cambina's "rod of peace " entwined with
two wedded serpents, "with one olive garland crowned."
This was probably emblematic of the impending reconcilia-
tion of the combatant heroes and their simultaneous espousals.
For " the rod which Maia's son doth wield, Wherewith the
hellish fiends he doth confound," which Spenser himself
introduces as resembling that borne by the lovely peace-
maker of his poem—was also snake bound ; and one
legend (though another less pleasing attaches to the staff of
Mercury) runs that Hermes once found two snakes fighting,
and, having separated them, twisted them round his *caduceus*,
or herald's staff, as typical of peace restored.   At first this
staff was of entwisted olive branches adorned with white
ribands—which is still the colour of peace ; but in later
representations of the herald divinity snakes take the place
of the ribands.   For a different reason the wand of "the

blameless physician " carried serpents, the art of Æsculapius being here symbolised by the creature which " renews its youth," and was supposed to have an instinctive knowledge of the healing virtues of herbs. The brazen serpent of the Mosaic wilderness had, in this restorative and curative significance, been anticipated in the temple courts of Epidaurus. It affords a simile on empiric nostrums. Thus Green, in his poem on *The Spleen*—

> " A corporation,
> The brazen serpent of the nation,
> Which, when hard accidents distressed,
> The poor must look at to be blessed."

But as a rule it is "vengeful;" "pernicious;" "with venom fraught;" "painted and empoisoned;" the supreme peril ("you might as safely waken a serpent," is an accepted comparison for the most hazardous enterprises [1]); a creature of secret ways, "more hid than paths of snakes" (Davenant); the uttermost symbol of desolation. "Palmyra's ruins " have no tenant but the hissing serpent " (Moore); it sits on the "Rajah's throne" when the lawful dynasty is extinct (Hemans); "rolls" through the "deserted market and the pleading-place, Choked with brambles and o'er-grown with grass " (Cowley): and so in Coleridge :—

> " The mighty columns were but sand,
> And lazy snakes trail o'er the level ruins."

The movement of the snake, so suggestive at once of subtlety and of strength, so wonderfully elegant and yet awe-inspiring, could not fail to arrest the poet's attention and provoke his admiration—

---

[1] " Pauses ere he wake
The slumbering venom of the folded snake :
The first may turn, but not avenge the blow,
The last expires, but leaves no living foe ;
Fast to the doom'd offender's form it clings,
And he may crush not conquer still it stings ! "—*Byron.*

D

> " And when they paint the serpent's scaly pride,
>   Their lines should hiss, their numbers smoothly glide."

Abundant recognition, therefore, is given to the striking grace with which these fearsome things sinuously glide, and as if obeying the attraction of some invisible magnet rather than progressing by any voluntary exercise of muscle, move from place to place. Many fine images are thus suggested, and finest of all is Keats :—

> " At this, through all his bulk an agony
>   Crept gradual, from the feet into the crown ;
>   Like a lithe serpent vast and muscular
>   Making slow way, with head and neck convulsed
>   From overstrained might."

But Virgil's fine picture of the wounded snake that "drags its slow length along" has many admirers :—

> " The trodden serpent on the grass
>   Long behind his wounded volume trails."

That wonderfully poetical touch in Nature of placing serpents in all her Edens, giving them the most exquisite foliage and flowers for their ambush, is not wasted on poets. But I cannot help thinking that they strike a false note when they make the presence of the snake detract from the beauty of the blossoms under which it hides. Roses, as Shakspeare says, are not "deceitful" because the adder is beneath. The contrast is itself sufficient, and if any moral is to be drawn, it might better be one of apology for the dangerous reptile in seeking such a resting-place than of reproach for the innocent flower. The rose, curiously enough, is specially selected as the serpent's retreat :—

> " As poisonous serpents make their dread repose
>   Beneath the covert of the fragrant rose."

Yet it is improbable that snakes often go to sleep under a rose bush, except our own harmless reptile in England. To tread on a serpent is proverbially perilous, but how ludicrous the whole situation becomes, when as in King, trodden gives place to *spurned*.

> " Fell as a spurned serpent as she shoots along
>   With lightning in her eyes, poison in her tongue."

Moreover, oddly enough, in plant-lore, this particular flower is one of those said to be distasteful to venomous reptiles.

That serpents specially affected the rose as their lurking-place, is, then, a poetical fancy, natural enough when the desire for strong contrast is needed, but not supported by any traditions. The beautiful, sweet-scented rose, the delight of the fair, is a striking contradiction to the "hideous, foul-smelling" reptile, the terror of the sex; and folk-lore does not encourage the association.

Snakes, so the people's traditions say, love lavender and walnut-trees and fennel. "More pleased my sense," says Satan to Eve, "than smell of sweetest fennel." It was supposed to assist them in casting their skins, thus restoring youth, and in brightening their eyes dimmed by old age. They hate and flee from hemlock, southernwood, and rue. Indeed, so violent and notorious is the reptile's aversion to the last-named that its antagonists take advantage of it, to fortify themselves against its assaults. Thus "when the Weesil is to fight with the Serpent, she armeth herselfe by eating Rue against the might of the Serpent;" which is a curious reproduction of the fiction of the mongoose's eating of the "aristolochia," another of the snake's vegetable anti-pathies.

As to the southernwood, I have my doubts, in spite of Lucan :—

> " There the large branches of the long-lived hart,
>   With southernwood their odours strong impart ;
>   The monsters of the land, the serpents fell,
>   Fly far away, and shun the hostile smell."

For during the Afghan campaign I rode through leagues

of it, and found snakes as common amongst it as in the
highly-aromatic sage-brush of Western America.   It is worth
noting, though, how, in a way analogous to the "doctrine
of signatures," the strong-smelling, aromatic snakes are
popularly connected with the most odorous herbs—lavender,
walnut, fennel, rue, and "old-man."   Nor will snakes, tradition
says, come under the juniper or the ash-tree.   So Cowley
has—

> " But that which gave more wonder than the rest,
>   Within an ash a serpent built her nest
>   And laid her eggs : where erst to come beneath
>   The very shadow of the ash were death."

"The leaves of this tree," saith an old herbalist, "are of so
great virtue against serpents that they dare not so much as
touch the morning and evening shadowes of the tree, but
shun them afar off, as Pliny reports.   He also affirmeth that
the serpent being hemmed in with boughes laid round about,
will sooner run into the fire, if any be there, than come neare
the boughes of the ash ; and that the ash flowereth befor
the serpents appear, and doth not cast its leaves befor
they be gon again.   'We write (saith he) upon experience,
that if the serpent be set within a circle of fire and the
branches, the serpent will sooner run into the fire than into
the boughes.   It is a wonderfull courtesie in nature, that
the ash should flower befor the serpents appear, and not
cast her leaves befor they be gon again.'"

If they have to be driven away, nothing is more efficacious
than the strewing of leaves of star-wort on the ground, or,
"which doth astonish them," sprigs of that virtuous herb
dittany.   Should any one be bitten by snakes, tradition
assures a complete antidote in the adder's-tongue fern :—

> " For them that are with newts, or snakes, or adders stung
>   He seeketh out an herb that is called adder's tongue ;
>   As Nature it ordained its own like hurt to cure,
>   And sportive did herself to niceties insure."

Are more wanted? Then take bramble leaves or herb-william, bugloss, horehound, betony, hawk-weed, or a cross made of hazel twigs. Indeed, bugloss and dittany will not only cure the bitten but kill the biter—all of which is very curious and pathetically human, seeing that these weeds are but common wayside wildings, and not some upas or man-chineel or dreadful Chilian serpent-tree. That these should chill the fiery, blood-kindling venom of snakes we could almost be content to imagine. Their own juices are too fatal for serpents' rivalry.

It is easy, therefore, to see how these reptiles came to possess the reputation of being cunning in herbs, and so, illogically (after the manner of popular beliefs), of being themselves medicinal, their flesh not only wholesome and curative but miraculous in its virtues, endowing with the knowledge of the speech of animals, and of the hiding-place of buried treasures, and their effigies to be the acknowledged crest and trade-mark of physicians from Æsculapius to Holloway. Hygeia herself always carries a serpent; and in this connection how delightfully consular and Roman is that anecdote of Exagoras, the ambassador from Cyprus. He came to Rome and bored them all so dreadfully with talking about the virtues of herbs and snakes that the consuls had him put into a tank full of serpents to test his long-winded theories. And the odd thing was, the vipers would not touch the ambassador.

Nor, as a testimonial to the serpent's ability as an herbalist, is the following incident to be neglected. Glaucus, son of Minos, died, and the king, his father, in a high-handed fashion, shut up a certain one in the family vault with the corpse, telling him that he should never come out alive unless his son did so too. The unfortunate man sat him down, disconsolately enough, we may suppose, by the side of the dead body, when, suddenly, there appeared a snake, which, as he saw it was about to crawl upon the bier, he killed.   Soon after, in came a second snake, and it, on perceiving its dead relative, hastily retreated, but, by-and-by, returning with a sprig of herb in its mouth, restored the first snake to life.   Acting on the hint, the prisoner took the precious vegetable, and rubbing Glaucus' corpse with it, had shortly the satisfaction of walking out of the mausoleum arm-in-arm with the revivified prince.

That the "glossy vine" was a "serpent charm" is a poetical tradition new to me, and may perhaps be an error for that other plant of Bacchus the "glossy" ivy.   Nor can I trace Wyatt's superstition

> "That snakes have twice to cast away their stings."

British snakes being harmless are not in harmony with the poetical serpent "idea," and seldom occur in verse.   In Coleridge and Wordsworth, "hurrying along the drifted forest leaves, the scared snake rustles;" but twice only, as far as I am aware, that pretty creature, our "grass-snake," receives kindly mention.   These notable exceptions are in the verse of that delightful poet of nature, Hurdis, who speaks in the Favourite Village of "the viper and the basking eft "—

> "And spotted snakes, innocuous as they glide,
>   With whisper not unwelcome,"

and Marvell, who presents his beloved shepherdess with "a harmless snake."

CHAPTER IV.

SNAKES IN TRADITION.

TRADITIONS, whether ancient or modern, all conspire to make the serpent-folk inhabitants either of a subterranean darkness or of the earth's most desolate places,—black creviced rocks and rotting vegetation, blistering desert sands and festering swamps. They are the outlaws of animal society. Erebus and lower Orcus and Tartarus below Hades know them. They are familiars of the gloomy shades by Styx, in the caverned banks of Acheron and Cocytus, the Cimmerian darkness beside Avernus. And wherever we find them, they are the rejected of creation, and for ever grovelling upon their bellies and sulkily tracing upon the dust the hieroglyphic record of the original curse.

Yet how differently the lives of these splendid and powerful beings are really passed. What creatures revel in more exquisite vegetation of leaf and blossom than the boas, anacondas, and pythons? and do not snakes share with fish their abodes in sea and river and lake? Indeed, there is no family of wild life that traverses so completely every experience of delightful habitation.

Nor does tradition sufficiently set forth the great snake-parable, with its awful significances of latent mischief, ambushed in such beauty. "Not even the plumage of the Birds of Paradise can excel the purples, blues, and gold of a python that has just cast its slough, while an infinite and terrible interest underlies those iridescent charms from the fact that the coils, soft as rose leaves, and shot with colours like a dove's breast, can crush the life out of a jaguar in all its rage, and slowly squeeze it into pulp. Watch its breathing; it is as gentle as a child's. Let danger threaten, however, and lightning is hardly quicker than the dart of those vengeful convolutions. The gleaming length rustles proudly into menace, and instead of the voluptuous lazy thing of a moment ago, the python, with all its terrors complete, erects itself defiantly, thrilling, so it seems, with eager passion in every scale, and tracing on the air with threatening head the circle within which is death." No wonder that the world has always held the serpent in awe, and that nations should have worshipped, and still worship, this emblem of destruction. It is fate itself, inevitable as destiny, deliberate as reason, incomprehensible as Providence. Yet in poetry they figure invariably as the instruments of divine wrath, the objects of popular detestation, the most hateful metamorphoses of humanity, the incarnations of sin. Their graces are deceits, their powers malign. From their very criminality they command reverence as being potential. Even the legends of their beneficence do them no good. They are wise, but only as the bad, as witches, as the devils, are wise. Humanity begrudges them even the credit for their lapses into benignity, and hardly forgives them honourable memories.

"Both gods and heroes alike held victory over the snake as the supreme criterion of valour. They graduated to divinity by slaying serpents. Indra and Vishnu conquer snakes, Hercules has his hydra, St. George his dragon, and Apollo his python. It is over the body of Ladon, terrible progeny of a terrible parentage—Typhon the father and Echidna the dam—that the hero steps to gather the golden apples; and across the dread chameleonising coils of Fafnir, that Sigurd reaches out his hand to the treasures of Brunhild on the glistening heath. What more fearful in Oriental myth than Vritna; the endless thing which the gods overcome; or

Kalinak, the black death ; or Ahi, the throttler ? Jason and Perseus, Feridun and Odin, claim triumph over the snake as the chiefest of their glories, and it would be tedious to recapitulate the multitude of myths through which 'the dire worm' has come down to our own time dignified and made awful by the honours and fears of the past."

Foremost of all the reptiles of tradition is that "spirited sly snake," "the enemy of mankind," that "unparadised the world."

> " Say first what cause
> Mov'd our grand parents, in that happy state,
> Favour'd of Heav'n so highly, to fall off
> From their Creator, and transgress His will
> For one restraint, lords of the world besides ?
> Who first seduc'd them to that foul revolt ?
> *Th' infernal Serpent.*"

In " Paradise Lost," the Satanic vehicle is always of great dignity, and however seriously it may be denounced is treated with severe respect. In some other poets it is scarcely so important a personage. Cowper, indeed, makes it ridiculous. Marvell has this quaint fancy—

> " When our first parents Paradise did grace,
> The serpent was the prelate of the place ;
> Fond Eve did, for this subtle tempter's sake,
> From the forbidden tree the pippin take ;
> His God and Lord this preacher did betray,
> To have the weaker vessel made his prey."

And Cowley this—

> " Oh Solitude ! first state of human kind,
> Which blessed remained till man did find
> Eve his own help and company ;
> As soon as two, alas ! together joined,
> The serpent made up three."

But with Milton my only fault is that he sat down to write of the Temptation unfairly prejudiced against the snake. He is fanatically, Puritanically, inflexible : refuses to give it, as he does King Charles, the benefit of a generous doubt. Neither the one nor the other was without compulsion in

error—but what of that? Like the stout Roundhead that
he was, he looked only at the ultimate offence, and would
not take extenuating circumstances into consideration—

> " The serpent sly
> Insinuating, wove with Gordian twine
> His braided train, and of his fatal guile
> Gave proof unheeded."

This passage is from Milton's description of the Garden
of Eden, and is his first reference to the creature which he
afterwards, when its body had been invaded by Satan, loads
with such infamy. It is for my purpose, a notable passage,
as showing how the great poet allowed his knowledge of the
sequel to prejudice him beforehand against the snake when
it was then, as he himself later on proves, a harmless beast,
a favourite of Eve's, and as yet perfectly innocent. Sin had
not as yet entered Paradise.

> " Frisking played
> All beasts of the earth, since wild, and of all chase
> In wood or wilderness, forest or den :
> Sporting the lion ramped, and in his paw
> Dandled the kid ; bears, tigers, ounces, pards,
> Gambol'd before them."

Only the snake is described as created in original sin and
being naturally vicious, of "fatal guile." Milton no doubt
thought he had inspired authority for separating this one
animal from all the others by such a terrific chasm, for in
Genesis we read, "Now, the serpent was more subtle than
any beast of the field." But I understand the authorities
upon the Scriptures would read this in an esoteric sense,
just as in the New Testament they do not take categorically
our Lord's advice, "Be wise as serpents." Was it really
intended by the Inspirer of Genesis that the snake should
be held up to calumny? At any rate, it seems to me incon-
testable that no poet had the privilege to "mar creation's
plan," by supposing that in the sinless garden there was

placed one sinful beast. The idea of this solitary iniquity in Paradise is intolerable.

Later on, Satan determines to pervert the snake, take possession of its body, and bedevil its innocent animal intelligence with something worse than human wickedness. He makes up his mind to find

> " The serpent sleeping, in whose mazy folds
>   To hide him, and the dark intent he brought ; "

and so descends in the form of a black mist to look for the devoted creature :—

> " Him fast-sleeping soon he found,
> In labyrinth of many a round self-rolled,
> His head the midst, well stor'd with subtle wiles ;
> Not yet in horrid shade or dismal den,
> Nor nocent yet, but on the grassy herb
> Fearless, unfeared, he slept. In at his mouth
> The Devil entered, and his brutal sense,
> In heart or head, possessing, soon inspired
> With act intelligential ; but his sleep
> Disturbed not."

Up to this point, therefore, the wretched serpent is the passive victim of a most atrocious trespass. Henceforth it is not its own self but " possessed," and no more to blame than the bedevilled swine of Gadara. It has been made the instrument of a designing villain ; which was its misfortune, not its fault ; and in its second state it was not, to my thinking, a bit more culpable than in its first. For it was not responsible for itself, being under the direct control of the Fiend " incarnate and imbruted " in its form. That the Creator subsequently judged otherwise, and took away the serpent's legs as a punishment for the part it had played in the great tragedy, only shows the infirmity of human judgment, and must be accepted in the same humility of mind as the visiting of a man's sins upon the fourth generation of his posterity, and quite apart from mortal theories of justice. Milton, however, would remove the apparent hardship of the serpent's lot—first misused by the Devil and then punished by God—by making it a conscious accomplice of Satan. He commences with saying that when everything else created was innocuous and amiable, it alone was filled with "fatal guile." Then when the Tempter finds it asleep, the poet describes its head as "well stored with subtle wiles;" and subsequently, when Eve hears it begin to talk, she addresses it in amazement thus :—

> " Thee, serpent, subtlest beast of all the field
> I knew, but not with human voice endued."

So that not only is the snake originally wicked, but Eve, that miracle of heavenly innocence, actually knows it. Surely this idea, that *suspicion* was present in Paradise, spoils the whole picture.

That the poet himself seems to recognise his difficulty is, I think, evident ; for besides his iteration of the original, native, badness of the serpent (itself significant), he makes Satan, when informing the chiefs of hell of his triumph over man, and the subsequent curse, deride thus :—

> " Me also hath He judged, *or rather*
> *Me not, but the brute serpent*, in whose shape
> Man I deceived."

Throughout the latter part of his speech Satan tries, and successfully, to make the Fall ridiculous, for his audience laugh when they hear about the apple ; and then he goes on to deride what seems to him, and to Milton, the vicarious culpation of the serpent.

However, to continue the poet's splendidly original description of the snake, we find Satan, a "mere serpent in appearance," searching all the favourite haunts of our first parents, and at last, "beyond his hope," he spies Eve all alone tending her flowers. The sight of her beauty strikes him at first "stupidly good," as the poet puts it, but immediately thereafter inflames him with fiercer envy of

Adam's happy lot, and he moves on, soliloquising as he goes upon the ruin he is about to work :—

> " So spake th' enemy of mankind, inclosed
> In serpent, inmate bad, and toward Eve
> Addressed his way, not with indented wave,
> Prone to the ground, as since, but on his rear,
> Circular base of rising folds, that towered
> Fold above fold a surging maze, his head
> Crested aloft, and carbuncle his eyes ;
> With burnished neck of verdant gold, erect
> Amidst his circling spires, that on the grass
> Floated redundant : pleasing was his shape,
> And lovely ; never since of serpent kind
> Lovelier."

But he does not approach Eve directly, but "with tract oblique, as one who sought access, but feared to interrupt, sidelong he worked his way," and when in her sight he displays himself to catch her attention :—

> " Curled many a wanton wreath in sight of Eve,
> To lure her eye."

She hears him rustling, but does not look up, being so accustomed to the beasts disporting themselves about her and vying with each other for her regard. So the snake comes right in front of her "as in gaze admiring."

> " Oft he bow'd
> His turret crest and sleek enamel'd neck,
> Fawning, and licked the ground whereon she trod.
> His gentle dumb expression turned at length
> The eye of Eve to mark his play."

Her attention thus won, the tragedy commences. She asks him in astonishment how he came to have human speech. He replies ("the spirited, sly snake"), by eating of a certain fruit. She asks where the tree stands ?—

> " To whom the wily adder, blithe and glad :
> Empress, the way is ready, and not long ;
> . . . If thou accept
> My conduct, I can bring thee thither soon."

"Lead, then," says Eve ; and the serpent willingly starts off.

> " He leading swiftly rolled
> In tangles, and made intricate seem straight,
> To mischief swift."

His crest flashes with hope, and like an *ignis fatuus* "glistered the dire snake" as he led Eve,

> " Our credulous mother, to the tree
> Of prohibition, root of all our woes."

She sees, is tempted, and falls.

> " Earth felt the wound ; and Nature from her seat
> Sighing, through all her works gave signs of woe,
> That all was lost !   Back to the thicket slunk
> The guilty serpent,"

and thereafter disappears from Eden.   The curse is pronounced, and Satan, reaching his own dominions, seats himself upon his throne, and addressing the assembly of fallen angels, "Thrones, dominations, princedoms, virtues, powers," boasts of what he has done, and then pauses for "their universal shout and high applause to fill his ear." Instead of that—

> " he hears
> On all sides, from innumerable tongues,
> A dismal universal hiss ; "

and then Satan begins to feel himself turning into a snake. " His arms clung to his ribs, his legs intwining each other," till he falls off his throne "a monstrous serpent, on his belly prone."   In vain he attempts to address his captains, for he can only hiss ; and then issuing from the hall the rout of "complicated monsters" swarm into the open air, where all the fallen host are awaiting their appearance, and, instead of their chiefs, see "a crowd of ugly serpents."   At the sight, horror seizes them and they begin to change too, "the dire form catched by contagion," until the whole of Satan's followers are turned to hissing snakes.   And lo ! by

divine miracle a grove of trees resembling the "dread probationary Tree" of Eden, heavy with luscious fruit; and
the scaly multitude, "rolling in heaps," scale the boughs,
hoping to eat. But the fruit turns to "bitter ashes" in
their mouths; yet goaded on by thirst and hunger, they
attempt to eat again and again; "with hatefullest disrelish
writhed their jaws, with soot and cinders filled;" then worn
out with famine and with "ceaseless hissing," they are
temporarily respited and resume their proper shapes; to
which Milton adds this legend :—

> " Yearly enjoined, some say, to undergo
>   This annual humbling certain numbered days,
>   To dash their pride, and joy for man seduced."

Such is the demoniacal serpent of Milton, and it is assuredly
a fine creation—the foremost reptile in poetry.

How pitifully inadequate, after such a dignified flight, is
Cowper's "flittermouse wing" attempting the same lofty
theme. Here, for instance, is his curse, a travesty upon
the original :—

> " Prone on thy belly, serpent, thou shalt grovel,
>   As if to man suggesting,
>   Dark as the riddling God, man is of clay;
>   And clay shalt thou be, destitute of soul,
>   As destitute of soul each other reptile."

It is the "Stygian," the "cruel" serpent, recognised by
both Adam and Eve as "empoisoned." Yet she admires
it, the monstrous hybrid :—

> "A human breast it has,
>   The rest is serpent all;
>   Oh! how the sun, emblazing with its rays
>   These gorgeous scales with glowing colours bright,
>   O'erwhelms my dazzled eyes."

And Adam specially points it out to her as a solace in
occasional solitude :—

> " If weary amidst the flowers,
>   Thou seek'st to close thine eyes,
>   Behold ! with flattering pinions at thy feet,
>   A serpent midst the flowers darts and hisses."

And can anything be more odious in comparison with the noble purity of Milton's treatment of the theme, than Cowper making Eve longingly guess at the pleasures of wedded life from observing the connubial complexities of snakes ?

> " Look there and see amidst the thousand folds
>   Those close entwisted snakes,
>   That in a single being seem combined—
>   Coy Adam, even these
>   Weave the close web of love."

This poet further makes Eve call the serpent to its face " snaky : "—

> " Your looks are snaky, and your glance malign."

As for Satan, whom Cowper calls " Beelzebub," he is a perfectly ridiculous personage, the very Bumble of devils.

In his poem on the Tree of Knowledge, Cowley supposes the serpent to be Pride, allegorically typified :—

> " Henceforth, said God, the wretched sons of earth
>   Shall sweat for food in vain,
>   That will not long sustain ;
>   And bring, with labour, forth each fond abortive birth ;
>   That serpent, too, their pride,
>   Which aims at things deny'd—
>   That learn'd and el'quent lust—
>   Instead of mounting high, shall creep upon the dust."

In the " Davideis " the snake again appears as the author of mischief.   The scene opens in Hell with Lucifer, in the form of a gigantic serpent, seated on high :—

> " Anon, a thousand devils run roaming in,
>   Some with a dreadful smile, deform'dly grin ;
>   Some stamp their cloven paws, some foam and tear
>   The gaping snakes from their black knotted hair."

But Lucifer is outrageously grotesque. The rising star of David, and the promise of the Messiah through his seed, has filled Satan with fury. He cannot conceal his emotion on hearing the news :—

> " Thrice did he knock his iron teeth, thrice howl ; "

and then, relieved, he reproaches his devils, and asks them why they are not up and doing, instead of stopping at home in Hell "playing with their idle serpents." But his rage chokes further utterance, and breaks off in a hemistich, short by two feet of the proper line, and

> " With that, with his long tail he lashed his breast,
> And, horribly, spoke out in looks the rest."

This outburst frightens the crowd into such silence that

> " No hiss of snake, no clank of chain, was known,
> The souls, amidst their tortures, durst not groan."

You might have heard a pin drop. But soon there is a movement in the "dire throng," and Envy crawls forth :—

> " Her black locks hung long,
> Attired with curling serpents ; her pale skin
> Was almost dropp'd from the sharp bones within ;
> And at her breast stuck vipers, which did prey
> Upon her panting heart."

This dreadful personage volunteers her services, and after recounting her previous exploits—very much after the manner of one of Mayne Reid's "braves" going out to scalp a paleface—offers to inspire Saul with mortal jealousy of David :—

> " She spoke. All stared at first, and made a pause ;
> But straight the general murmur of applause
> Ran through Death's courts."

"Great Beelzebub" starts up "to embrace the fiend," but she dodges him and is off.

> " The snakes all hissed, the fiends all murmured."

Arriving at Saul's palace, she assumes the form of "Father
Benjamin," and approaching the slumbering prince, upbraids
him for allowing David, "a boy and minstrel," to steal away
his people's love and his crown, and exhorts him to be
"whole Saul," and rid himself of the son of Jesse.

> " With that she takes
> One of her worst, her best-beloved, snakes ;
> ' Softly, dear worm ! soft and unseen,' said she,
> ' Into his bosom steal, and in it be
> My viceroy ! ' "

Cowley, again, speaking of the transformation of Aaron's
rod (which he calls " Moses' wand ") says :—

> " It gaped and hissed aloud,
> With flaming eyes survey'd the trembling crowd,
> And like a basilisk almost looked the assembly dead.
> Swift fled th' amazed king, the guards before him fled ; "

which is a curious misreading of Holy Writ, inasmuch
as Pharaoh, in the hardness of his heart, only recognised
in the miracle one of the commonest tricks of his own court
jugglers. Instead of flying amazed before "the Almighty
wand," he sent, we are told, for the "magicians of Egypt,"
who, at his orders, repeated the miracle. Cowley's version
of the incident is as follows :—

> " Jannes and Jambres stopp'd their fight,
> And with proud words allay'd th' affright.
> ' The god of slaves,' said they, ' how can he be
> More powerful than their master's deity ?'
> And down they cast their rods,
> And muttered secret sounds that charm the servile gods.
> The evil spirits their charms obey.
> All in a subtle cloud they snatched the rods away,
> And serpents in their place the airy jugglers lay.
> Serpents in Egypt's monstrous land
> Were ready still at hand,
> And all at th' Old Serpent's first command :
> And they, too, gaped, and they, too, hissed,
> And they their threatening tails did twist ;

> But straight on both the Hebrew serpent flew,
> Broke both their active backs, and both it slew,
> And both almost at once devoured."

But it will be seen that here the poet as much detracts from the authorised narrative as previously he had added to it, inasmuch as from Holy Writ we gather that Aaron's rod devoured more than two of the others, for the verse runs, "They cast down every man his rod and they became serpents: but Aaron's rod swallowed up their rods." For myself I confess I have always imagined—fancy being led thereto by a picture in an illustrated Bible which much attracted me in childhood—that the floor of Pharaoh's palace was fairly littered with snakes, and that no part of the miracle was quite so miraculous as Aaron's serpent—which, as depicted, was no bigger than most, and not so big as some—being able to contain all the rest. Doubtless my timid scepticism on this point was conciliated by being reminded of the extraordinary containing-capacity of other kinds of snakes, and by some such zoological "fact" as the boa-constrictor's habit of swallowing bulls.

Crashaw has exactly the same story as that of the "Davideis," except that Cruelty takes the place of Envy. Lucifer, "mischief's old Master," hears in Hell of the birth of the Messiah, and determines to baulk the Almighty. He summons his ministers, and, though many volunteer, he selects Cruelty—whom the poet calls "the fourth Fury"—and sends her forth to pervert Herod. Assuming the shape of Father Joseph, she approaches the sleeping Tetrarch, and, in language very similar indeed to that of the "Davideis," fills his breast with the horrible suggestion of the Massacre of the Innocents. "Be Herod," she whispers, and vanishes.

> " So said her richest snake, which to her wrist,
> For a beseeming bracelet she had tied,
> (A special worm it was, as ever kissed
> The foamy lips of Cerberus) she applied

> To the king's heart ; the snake no sooner hissed
> But Virtue heard it, and away she hied."

His Hell, by the way, is simply a wilderness of snakes.

How grandly different are Milton's pictures of the Arch-Fiend in his monster-shape ! The scene is the Tartarian lake, "as one great furnace flaming," a "fiery deluge, fed with ever-burning sulphur unconsumed." Satan recognises Beelzebub "weltering by his side," and they converse. The Arch-Fiend is thus described :—

> " With head uplift above the wave, and eyes
> That sparkling blazed, his other parts beside
> Prone on the flood, extended long and large,
> Lay floating many a rood, in bulk as large
> As whom the fables name of monstrous size,
> Titanian, or Earth-born, that warr'd on Jove,
> Briareus or Typhon, whom the sea
> By ancient Tarsus held, or that sea-beast,
> Leviathan, which God, of all His works,
> Created hugest that swim th' ocean stream."

Later, when he is in eager flight, this fine image is employed :—

> " As when a gyphon through the wilderness,
> With wingèd course, o'er hill or moor or dale,
> Pursues the Arimaspian, who by stealth
> Had from his wakeful custody purloined
> The guarded gold."

And afterwards, when the actual transformation of the rebellious host into serpents is described, we see their chief

> " Still greatest in the midst,
> Now dragon grown, larger than whom the sun
> Ingender'd in the Pythian vale or slime,
> Huge Python." [1]

There is nobility of fancy throughout.

Traditionally the snake is "crested," so no poet refers to a snake of any importance without mentioning its crest.

[1] The python, curiously enough, has the vestiges of legs.—P. R.

This is a poetical rule to which there is no exception. It is Biblical, Homeric, heraldic, but none the less preposterous. No snake has a crest. Some have inflations of the neck below the head: a few very small vipers have prickles upon their heads. But there is nothing in all herpetology to warrant the "koruthaiolos" idea in which poets delight. Classical tradition, however, abounds with it. So Milton adopts it, and all others follow his example. But this and other varieties of the poetic "basilisk" will be found duly treated of in their proper places among the Fauna of Fancy.

Among the individual snakes of tradition, foremost, perhaps, are those of the Furies—

> " Revenge ! Revenge ! Timotheus cries ;
> See the furies arise ! · .
> See the snakes that they rear,
> How they hiss in their hair !
> And the sparkles that flash
>> From their eyes ! "

—more ancient than the Olympian gods, living in dark Tartarus, but issuing thence to punish the infamous with perpetual unrest and successive miseries. Tisiphone, too—

> "A hundred snakes her gloomy visage shade,
> A hundred serpents guard her horrid head."

and "fierce Alecto," with snaky tresses that listen and watch while she sleeps ; and Megæra—

> " Tossing her vipers round,
> Which, hissing, pour their poison on the ground."

The Gorgon again, "terrible Medusa," with her "long snaky locks of adder-black hair"—

> " His dark hair
> That pale brow wildly wreathing round,
> As if the Gorgon there had bound
> The sablest of the serpent-braid
> That o'er her fearful forehead stray'd."—*Byron.*

Surely one of the saddest, most unfortunate of maidens,

once so beautiful that gods wooed her, afterwards so dreadful that the mere sight of her face terrified men into stone —a notable illustration of "disastrous love" and of the malignity of female revenge. Temples, no doubt, ought to be respected, but whenever I think of Medusa's fate and her "fearful head" with its "crested" snakes, my opinion of Poseidon is not complimentary to that amphibious divinity.

Next, the assailants of Laocoön, immortalised in noble epic and almost as noble marble :—

> " Round sire and sons the scaly monsters rolled
>   Ring above ring, in many a tangled fold ;
>   Close, and more close, their writhing limbs surround,
>   And fix with foaming teeth the envenomed wound."

> " Or, turning to the Vatican, go see
>   Laocoön's torture dignifying pain—
>   A father's love and mortal's agony
>   With an immortal's patience blending :—Vain
>   The struggle ; vain, against the coiling strain
>   And gripe, and deepening of the dragon's grasp,
>   The old man's clench ; the long envenom'd chain
>   Rivets the living links,—the enormous asp
>   Enforces pang on pang, and stifles gasp on gasp."—*Byron.*

> " At last her utmost Masterpieces she found,
>   That Maro fir'd ; the miserable sire,
>   Wrapt with his sons in Fate's severest grasp.
>   The serpents, twisting round, their stringent folds
>   Inextricable tie. Such passion here,
>   Such agonies, such bitterness of pain,
>   Seem so to tremble through the tortur'd stone,
>   That the touch'd heart engrosses all the view.
>   Almost unmask'd the best proportions pass,
>   That ever Greece beheld ; and, seen alone,
>   On the rapt eye th' imperious passions seize :
>   The father's double pangs, both for himself
>   And sons convuls'd ; to Heaven his rueful look,
>   Imploring aid, and half-accusing, cast ;
>   His fell despair with indignation mixt,
>   As the strong-curling monsters from his side
>   His full-extended fury cannot tear.

> More tender touch'd, with varied art, his sons
> All the soft rage of younger passions show,
> In a boy's helpless fate one sinks oppress'd ;
> While, yet unpierc'd, the frighted other tries
> His foot to steal out of the horrid twine."—*Thomson.*

Then the "dread snakes," who "at Juno's vengeful nod, climbed round the cradle of the sleeping god." But the baby happens to be Hercules, and—

> " Waked by the shrilling hiss and rustling sound,
> And shrieks of fair attendants fainting round,
> Their gasping throats with clenching hands he holds,
> And death entwists their convoluted folds."

Of Cadmus, "how with the serpent's teeth he sowed the soil, and reaped an iron harvest of his toil!" from which Coleridge draws the moral :—

> " Who sows the serpent's teeth, let him not hope
> To reap a joyous harvest. Every crime
> Has, in the moment of its preparation,
> Its own avenging angel, dark misgiving,
> An ominous sinking at the inmost heart."

Of Iapetus, whom Keats sees grasping

> " A serpent's plashy neck, its barbèd tongue
> Squeezed from the gorge, and all its uncurl'd length
> Dead : and because the creature could not spit
> Its poison in the eyes of conquering Jove."

The living bolts of the warring Titans, of Apollo's prowess, "Latona's bane "—

> " Who slew Phiton the serpent where he lay
> Sleeping against the sun upon a day."

Of the Egyptian Cleopatra, "regal dressed, with the aspic at her breast "—

> " I am the worm the weary prize,
> The Nile's soft asp,
> One that a Queen has loved to clasp."

Nor are the serpent-reverences of contemporary cult forgotten, the snake-gods of India, where, as Sir W. Jones
sings—

> " Taraka, with snaky legions,
>   Envious of supernal powers,
>   Menaces old Meru's golden head,
>   And Indra's beamy regions,
>   With desolation wild bespread."

Foremost of these is Anantas, "the king of serpents,
with his thousand heads," the infinite—

> " That serpent old,
>   Which clasped the great world in its fold,
>   And brooded over earth and the charmed sea
>   Like endless, restless, drear Eternity;"

and next Shesh, "whose diamond sun makes subterranean
day." The poet refers here without a doubt to that fine
legend of the Indian aborigines the Nagas, or "snake-men,"
who say that once upon a time, and perhaps they are right,
they possessed the land, but were driven into the hill-fast-
nesses which they now inhabit by successive waves of
invasion, and that their great captain and divinity—Shesh,
"the king of serpents"—fled underground, and in contempt of the sunlight from which he had been exiled,
created the Kanthi-stone, more brilliant than a whole rock
of diamond, by the light of which he keeps the diary
of the earth, and solemnly records the procession of the
ages.[1]

This Shesh, "that never dies," and "whose hiss the round

---

[1] The Cherokee Indians of the West have much the same legend as the
Nagas of the East, and Mrs. Hemans refers to

> " The mighty serpent king
>   Midst the grey rocks, his old domain,"

who is supposed to dwell in the central recesses of the mountains, the chief
of the rattlesnakes, and who, though subterranean, is honoured as the
"light-giver."—P. R.

creation awes," is a reptile worthy of homage, and may be accepted without hesitation and in defiance of all sea-serpents, past and future, as the greatest snake on record. When Vishnu and the gods meet to extort from the sea the ichor of immortality, they pluck up from the Himalayan range the biggest mountain in it, and this they make their churn, while around it, as the strongest tackle they could think of, they bound the serpent Shesh. And the gods took hold of the head, and the devils took hold of the tail, and, alternately tugging, they made the mountain spin round and round, until the sea was churned into froth, and from the churning came up all the treasures of the deep, and the most precious possessions of man, and last of all Immortality. The gods and the devils scrambled for all the good things, but nothing more is said of the serpent who had been so useful, nor what he got for his services. Antiquaries in the West incline to think that he remained in the sea and became the kraken; but the Nagas believe him to be still under the hills dispensing fate by the light of a diamond.

When, too, Lakshmi fixes her admiring eyes upon "the azure Hari," he started at the summons of love :—

> " Straight o'er the deep, then dimpling smooth, he rushed,
> And towards th' unmeasured snake's stupendous bed
> The world's great mother, not reluctant, led.
> All Nature glow'd whene'er she smiled or blushed ;
>   The king of serpents hushed
> His thousand heads, where diamond mirrors blazed
> That multiplied her image as he gazed."

To these succeed a procession of the "serpents of Romance," "sleepless and stern to guard the golden sight;" the great reptiles of knightly story. Southey's " Romance's serpent winds the glittering fold." Of the victories of the chivalrous over these baleful monsters there is no end :—

> " Soon to a yawning rift Chance turn'd my way,
> A den it prov'd where a huge serpent lay ;
> Flame ey'd he lay :—he rages now for food,
> Meets my first glance, and meditates my blood.
> His bulk, in many a gather'd orb uproll'd,
> Rears spire on spire.   His scales bedropt with gold
> Shine burnish'd in the sun.   Such height they gain
> They dart green lustre on the distant main.
> Now wreath'd in dreadful slope, he stoops his crest,
> Furious to fix on my unshielded breast !
> Just as he springs my sabre smites the foe ;
> Headless he falls beneath the unerring blow.
> Wrath yet remains, though strength his fabric leaves,
> And the meant hiss the gaping mouth deceives ;
> 'The length'ning trunk slow loosens ev'ry fold,
> Lingers in life, then stretches stiff and cold."—*Savage.*

But, as characteristic of human anxiety never to leave triumph wholly with the reptile that was "devoted to defeat" in Eden, the following quotations (from Drayton and the "Reliques") showing how innocence can vanquish, even though unarmed, are well worth notice :—

> " Him by strength into a dungeon thrust,
> In whose black bottom, long two serpents had remain'd
> (Bred in the common sewer that all the city drain'd),
> Empoisoning with their smell ; which seized him for their prey ;
> With whom in struggling long (besmeared with blood and clay)
> He rent their squalid chaps, and from the prison 'scaped."

The following is even more to the point :—

> " And adders, snakes, and toads therein,
> As afterwards was known,
> Long in this loathsome vault had bin,
> And were to monsters grown."

Into this foul and fearful place the fair one, innocent, was cast.

> " The door being open'd straight they found
> The virgin stretch'd along ;
> Two dreadful snakes had wrapt her round,
> Which her to death had stung.

One round her legs, her thighs, her waist,
  Had twin'd his fatal wreath :
The other close her neck embrac'd,
  And stopt her gentle breath.

The snakes, being from her body thrust,
  Their bellies were so fill'd,
That with excess of blood they burst,
  Thus with their prey were kill'd."

But distinct from all others is the tradition of the Lamia—

" Thou smoothed-lipped serpent surely high-inspired,
  Thou beauteous wreath with melancholy eyes."

" Philostratus, in his fourth book, hath a memorable instance in this kind, which I may not omit, of one Menippus Lycius, a young man twenty-five years of age, that, going betwixt Cenchreas and Corinth, met such a phantasm in the habit of a fair gentlewoman, which, after taking him by the hand, carried him home to her house, in the suburbs of Corinth, and told him she was a Phœnician by birth, and if he would tarry with her he would hear her sing and play, and drink such wine as never any drank, and no man should molest him ; but she, being fair and lovely, would die with him, that was fair and lovely to behold. The young man, a philosopher, otherwise staid and discreet, able to moderate his passions, though not that of love, tarried with her awhile to his great content, and at last married her, to whose wedding, among other guests, came Apollonious, who, by some probable conjectures, found her out to be a serpent, a Lamia ; and that all her furniture was, like Tantalus' gold described by Homer, no substance, but mere illusions. When she saw herself descried she wept, and desired Apollonius to be silent, but he would not be moved, and thereupon she, plate, house, and all that was in it, vanished in an instant ; many thousands took notice of this fact, for it was done in the midst of Greece."

This tradition, a favourite with the poets, has its finest

exposition in Keats' deathless verse.   Mercury, roaming in
the pinewood in search of a nymph whom he loves, hears a
mournful voice bewailing itself, and searching among the
bushes finds

> " A palpitating snake,
> Bright and cirque-couchant, in a dusky brake ;
> She was a gordian shape of dazzling hue,
> Vermilion-spotted, golden, green, and blue ;
> Striped like a zebra, freckled like a pard,
> Eyed like a peacock, and all crimson barr'd ;
> And full of silver moons, that, as she breathed,
> Dissolved, or brighter shone, or interwreathed
> Their lustres with the gloomier tapestries—
> So rainbow-sided, touch'd with miseries,
> She seemed at once some penanced lady-elf,
> Some demon's mistress, or the demon's self.
> Upon her crest she wore a wannish fire,
> Sprinkled with stars, like Ariadne's tiar ;
> Her head was serpent, but ah, bitter sweet !
> She had a woman's mouth, with all its pearls complete.
> And for her eyes—what could such eyes do there
> But weep, and weep, that they were born so fair,
> As Proserpine still weeps for her Sicilian air ?
> Her throat was serpent, but the words she spake
> Came, as through bubbling honey, for love's sake."

He asks the cause of its woe, and the Lamia then says that if
Hermes will restore her to her human shape she will make
his love who is invisible appear before him.   The compact
is faithfully adhered to.   The god flies away with his
nymph into "the green-recessèd" woods, and the snake is
alone.

> " Left to herself, the serpent now began
> To change ; her elfin blood in madness ran ;
> Her mouth foam'd, and the grass, therewith besprent,
> Wither'd at dew so sweet and virulent ;
> Her eyes in torture fix'd, and anguish drear,
> Hot, glazed, and wide, with lid-lashes all sear,
> Flash'd phosphor and sharp sparks, without one cooling tear.
> The colours all inflamed throughout her train,
> She writhed about, convulsed with scarlet pain ;
> A deep volcanian yellow took the place
> Of all her milder-moonèd body's grace,

> And, as the lava ravishes the mead,
> Spoilt all her silver mail, and golden brede :
> Made gloom of all her frecklings, streaks, and bars,
> Eclipsed her crescents, and lick'd up her stars :
> So that, in moments few, she was undrest
> Of all her sapphires, gems, and amethyst."

Thereafter she vanishes, goes to Corinth, bewitches a youth of that city, who marries her. To the wedding feast, unbidden, comes Apollonius the sage, who detects her, fixes his eye, "keen, cruel, penetrant, stinging," upon her. She implores him by gesture to look away. The bridegroom beseeches the sage to spare her. "*Fool !*" says Apollonius to the one, and then to the other, "*Serpent !*" Upon this she gives a frightful scream and is gone—

> "And Lysius' arms were empty of delight,
> As were his limbs of life, from that same night."

CHAPTER V.

THE POETS' SNAKES.

In folklore the snake has, I find, three distinct and different aspects. Only one of these—the malignant aspect—is recognised in verse.

Elsewhere, however, it is very often met with as the faithful custodian of treasure, and nearly every country has its Serpent which guards the all-important tree of immortality and other secrets, or its dragonish thing that defends some priceless possession. The Greeks only fled from Athens when they heard that the snake of the city had deserted the Acropolis, and it is only a few years ago—in this very nineteenth century of ours—that the Nagas of India formally surrendered to British troops because their serpent oracle had escaped from its priests. For the snake is the universal guardian of the under world, whether we look for it in the diamond-lit caverns of Shesh in Hindostan, or under the ash Ygdrasil in Norway.

In its second aspect it is benign, and emblematical of providential wisdom and a vigilant solicitude. So we find it in Anantas, the infinite, lending itself to the gods, that they might use its body for a rope, to be tied round the mountain of Meru when they churned the ocean ; as the beneficent rainbow of Africa ; in the "feathered serpent" of South America that taught men religion and gave them the gift of wine ; in Hoa, the third person of the Babylonian trinity, that befriends the penitent.

Its third aspect, and the only one in which the poets regard it, is diabolical; typifying a malignant darkness that is hurtful to man, and symbolising every wicked mood or motive, every misery, in human nature.

"Slander's serpent mark," says one; and in Keats' tragedy of "Otho," the fair Auranthe and her brother who calumniate the Princess, are "those two vipers from whose jaws a deadly breath went forth to taint and blast the guileless lady."

Envy is very often the motive for slander, so this also is another of the "serpentine obliquities of life."

> " Envy, with serpent eye,
>   Marks each praise that soars on high."

It is "snake-hung," "hissing," and (from the mystic properties of serpents) " wizard " envy is armed with " venomed teeth." Personified as a hideous hag, it lives in gloomy dens and is " black envy," except when, as in the " Davideis," she is described as the direst fiend in Hell. Pride and envy are not unnaturally associated :—

> " From what cause can envy spring?
>   Or why embosom we a viper's sting?
>   'Tis envy stings our darling passion Pride."

Here the poetical diagnosis goes one step further back. A man's ideas of his own merit often make him resent the recognition of another's, whence, no doubt, slanderous depreciation. Being serpentine it is secret, whether lurking in some "cave," or cherished in the heart.

> " In his bosom secretly there lay
>   An hateful snake."

This, again, leads on to the fancy of man taking a sin to his heart for his own destruction, as in Æsop's fable of the Countryman and the Viper :—

> " He who in wild wood alleys roams, unthinking and unwise,
>   And takes a serpent to his heart for beauty of its eyes,
>   For splendour of its arching neck and glitter of its skin,
>   Was scarcely such a dupe as I, in ignorance of sin."

So Leyden's lines "cherished bosom-sin, Like nestling serpent gnawing the heart within," and "the green-eyed viper gnawing at my heart," or Shelley's

> " Foul and cruel thoughts, which feed
>   Upon the withering life within
>   Like vipers on some poisonous weed."

This is a curious passage, not only for the last line, which makes snakes vegetarian, but because the simile is so oddly at fault. The only association suggested, it seems to me, is between "withering" in the second line and "weed" in the third, but even that is too weak, while it is hardly sense to say that a thought, feeding on a withering life, resembles a viper eating a poisonous weed. The explanation, probably, is that the poet lapsed from his first image by what the wise call "some process of unconscious cerebration," his mind passing without intermediate expression in words from one to the other, and leaving, therefore, a gap without stepping-stones or bridge. In the first two lines Shelley, it may be, glances at a fiction which has never failed to attract poets, that of the snake living upon the vitals of its parents, as, for example, in Churchill, Dryden, and Marvell :—

> " Oh ! my poor country ! Devour'd
>   By vipers, which in thine own entrails bred,
>   Prey on thy life, and with thy blood are fed,
>   When children us'd their parents to dethrone,
>   And gnaw their way, like vipers, to the crown."

> " Against themselves their witnesses will swear,
>   Till, viper-like, their mother-plot they tear,
>   And suck for nutriment that bloody gore,
>   Which was their principle of life before."

> " These vipers have their mother's entrails torn,
>   And would by force a second time be born."

This idea, that the young vipers killed their mothers before coming into the world themselves, is one of great antiquity, and parricides, who were punished by the ancients by drowning, were sewn up in sacks, in which, as appropriate company for such criminals, some vipers had previously been put. Nor is the fiction by any means exploded, for in parts of England it is still believed that the adder—which brings forth its young alive, and does not, like the harmless "grass-snake," lay eggs—is killed by her progeny. How this happens rural superstition can have only the vaguest notion, for another, which contradicts it, namely, that the young adders always take refuge inside their mother if suddenly startled, exists simultaneously with it.

However, to return to the snakiness of human passions. Maternal anguish has her "torturing snakes." Remorse carries "a whip" (Southey) made of them, and can "dart poison through the conscious heart" (Akenside) : Superstition is (in Shelley) a " hundred-forked " snake "insatiate ; " Fierce debate (in Sackville) "deadly full of snaky heare ; " Conscience is " a stinging worm," with the " viper fear " (Green), and the " adder of disgrace " (Dodd) is " an undying serpent " calling (Shelley) " her venomous brood to their nocturnal task." Error—"serpent error wandering" of Milton—has a " poisonous serpent head ; " and in Cowper—

> " Sing, muse (if such a theme so dark, so long,
>   May find a muse to grace it with a song),
>   By what unseen and unsuspected arts
>   The serpent Error twines round human hearts,
>   Tell where she lurks, beneath what flowery shades,
>   That not a glimpse of genuine light pervades,
>   The poisoning, black, insinuating worm
>   Successfully conceals her loathsome form."

So, too, Pride the cause of error :—

> " Though various foes against the truth combine,
>   Pride above all opposes her design,

> Pride of growth superior to the rest,
> The subtlest serpent with the loftiest crest,
> Swells at the thought, and, kindling into rage,
> Would hiss the cherub Mercy from the stage."

Faith misplaced is in Shelley a "Python;" it is a curious passage :—

> " And Faith, the Python, undefeated,
> Even to its bloodstained altar-steps dragged on
> Her foul and wounded train, and men
> Were trampled and deceived again."

And elsewhere from another standpoint the poet has the same sentiment :—

> " Faith and Custom and low-thoughted cares
> Like thunder-stricken dragons, for a space
> Left the torn human heart, their food and dwelling-place."

"Faithlessness" has in Byron a serpent-fold—a fine idea : "to find in Hope but the renewed caress, the serpent-fold of further faithlessness"—and Coleridge associates the two :—

> " tyrants' promises
> That can enchant the serpent treachery
> From Faith, its lurking hole in the heart."

Jealousy is a "deaf adder :"—

> " The serpent of the field, by art
> And spell, is won from harming ;
> But that which coils around the heart,
> Oh ! who hath power of charming ?"

The "Scorner" has in Heber a "serpent tongue." Hate and lawless Pleasure are snakes, and

> " in the soul
> Lurks sin, the Serpent with the fiery sting
> Of sorrow, rankling in the conscience deep."

Avarice is a serpent (Savage), and so—"strange fellowship through mutual hate "—are Fear and Lust (Shelley). Care

(Churchill) and Evil (Akenside) are vipers, while Ben Jonson adds to the list, in the following powerful lines, the faithless, the selfish man :—

> " Look on the false and cunning man, that loves
> No person nor is loved : what ways he proves
> To gain upon his belly ; and at last
> Crushed into the snaky brakes that he had passed."

Love, of course, is a serpent, in a score or two of poets :—

> " A serpent nourish I under my wing,
> And now, of nature, 'ginneth it to sting."

Nor does the breaking of the heart cure, but rather aggravate the evil :—

> " I thought that this some remedy might prove ;
> But, oh ! the mighty serpent Love
> Broke, by this chance, in pieces small,
> In all still lived, and still it stung in all."

Nor are even "the loves of the Angels" secure from reptilian intrusion :—

> " When Love hath not a shrine so pure,
> So holy, but the serpent Sin,
> In moments even the most secure,
> Beneath his altar may glide in."

The jealousy of rivals, and the pleasure of inflicting pain, and the pain of suffering it, suggests the snake to many poets. Thus Wyatt in his "Jealous Man :"—

> " The wand'ring gadling in the summer tide
> That finds the adder with his reckless foot
> Starts not dismay'd so suddenly aside
> As jealous despite did, though there were no boot,
> When that he saw me sitting by her side
> That of my health is very crop and root.
> It pleased me then to have so fair a grace
> To sting the heart, that would have had my place."

Among the classes of individuals specially colubrine are

Critics (a particular aversion of poets), as in Goldsmith and
Crabbe :—

> " His wand's a modern author's pen ;
> The serpents round about it twined
> Denote him of the reptile kind,
> Denote the rage with which he writes,
> His frothy slaver venomed bites."

Or

> " The serpent Critic's rising hiss."

And Harlots—so when Dalila leaves the crippled hero's
presence with a taunt, the Chorus cry :—

> " She's gone, a manifest serpent by her sting,
> Discovered in the end, till now concealed ; "

and many others, down to the "unfortunate" of the London
streets, against whom Mackay launches this exhortation to
the "fool : "—

> " A serpent, woman headed,
> With loose and floating hair,
> Beware, O fool ! how you touch it,
> Beware for your soul !   Beware !
> 'Tis beautiful to look at
> As it rustles through the street,
> But its eyes, though bright as sunshine,
> Have the glow of hell's own heat.
>
> .　.　.　.　.　.　.　.
>
> 'Twill murmur soft sweet music
> To draw you to its mesh,
> And coil about you fondly
> To feed upon your flesh ;
> Beware of this flaunting Gorgon
> With the snakes in her wavy hair !
> Beware, O fool ! how you touch her,
> Beware for your soul, beware ! "

So, too, in Savage, Lawyers :—

> " Not a gay serpent glittering to the eye,
> But more than serpent or than harlot sly,
> For lawyer-like, a fiend no wit can 'scape,
> The demon stands confessed in human shape."

And in Coleridge's " Devil's Thoughts : "—

> " He saw a lawyer killing a viper
>   On a dung heap beside his stable,
>   And the devil smiled, for it put him in mind
>   Of Cain and his brother Abel."

Many families, alas! appreciate all too sadly Byron's picture of that insidious wretch who, either calling herself the friend of one or other, comes between husband and wife, or parent and children, and working mischief under pretence of impartial advice, widens the breach with a show of healing it :—

> " O serpent under femininitee
>   Like to the serpent depe in helle ybound :
>   O feined woman, all that may confound
>   Vertue and innocence, thurgh thy malice,
>   Is bred in thee, as nest of every vice."—*Chaucer.*

Or else as the friend of both, poisons each mutually against the other, carrying to and fro not the peace-making con- cessions, the timid beginnings of conciliation with which she is entrusted, but embittering hints and irritating suggestions that provoke recrimination, and further involve the already complicated difficulty.   She found a rift which a single kiss might have closed; she leaves a grief which passionate re- pentance, stretching from to-day to the tomb, cannot bridge.

> " She rules the circle which she served before ;
>   If mothers none know why before her quake ;
>   If daughters dread her for the mother's sake ;
>   If early habits those false links, which bind
>   At times the loftiest to the meanest mind
>   Have given her power too deeply to instil
>   The angry essence of her deadly will ;
>   If like a snake she steal within your walls,
>   Till the black slime betray her as she crawls ;
>   If like a viper to the heart she wind,
>   And leave the venom there she did not find,
>   What marvel that this hag of hatred works
>   Eternal evil latent as she lurks,
>   To make a Pandemonium where she dwells,
>   And reign the Hecate of domestic hells ?
>   Skill'd by a touch to deepen scandal's tints
>   With all the kind mendacity of hints,

> While mingling truth with falsehoods, sneers with smiles,
> A thread of candour with a web of wiles;
> A plain, blunt show of briefly-spoken seeming
> To hide her bloodless heart's soul-hardened scheming;
> A lip of lies, a face formed to conceal,
> And, without feeling, mock at all who feel."

So, too, in Cowper's "ancient prude"—suggested by Hogarth's picture of morning—we have Miss Bridget :—

> " Of temper as envenomed as an asp,
> Censorious, and her every word a wasp,"

the meddlesome, scandal-mongering, reputation-tearing old maid, who is all too familiar in society :—

> " Mark how the channels of her yellow blood
> Ooze to her skin and stagnate there to mud;
> Cased like the centipede in saffron mail,
> Or darker greenness of the scorpion's scale—
> For drawn from reptiles only may we trace
> Congenial colours in that soul or face."

Fever "like a serpent crawling," "insidious Ague, serpent-like," "Gout half a snake," and other ills that flesh is heir to, share in the serpent reproach; and both the north and the east wind, that cause so many of them in our chilly latitudes, have "serpents' fangs" and sting.

The image of the river " stealing like a silver snake," "a glistering snake," that "through the grassy mead, winds on, now hidden, glittering now in light" (Southey), is sufficiently hackneyed, but it suggests to Wordsworth this excellent additional fancy :—

> " A mightier river winds from realm to realm,
> And, like a serpent, shows his glittering back
> Bespotted with innumerable isles."

And to Shelley this :—

> " The glaciers creep
> Like snakes that watch their prey, from their far fountains
> Slow rolling on."

Several other points in the natural economy of the tribe

afford the poets incessant opportunity for metaphor.   Thus, the casting of their slough—

> " So may my spirit cast,
>   Serpent like, off the past."—*Herbert.*

> " What shall I do?   Where go?
>   When I have cast this serpent-skin of woe."—*Keats.*

> " But time strips our illusions of their hue,
>   And one by one in turn, some grand mistake
>   Casts off its bright skin."—*Byron.*

> " Warmed by health as serpents in the Spring
>   Aside their slough of indolence they fling."—*Crabbe.*

> " They've tricks to cast their sins
>   As easy as serpents do their skins
>   That in a while grow out again."—*Butler.*

That the snake leaves a slimy track is a popular error, already noticed as accepted by poets.   Slander and envy, therefore, are said to "befoul" as they go: the critic's "frothy slaver," the parasite's path "snake-marked," the traitor's "slimy clue," and other similar phrases have their source in the same fiction : "the trail of the serpent is over them all."

The image of the snake lurking in the herbage is as old as grass itself.   Our earliest poets, therefore, had it :—

> " I know under the green the serpent how he lurks ; "

and

> " Ware fro the serpent that so slily creepeth
>   Under the grass, and stingeth subtelly."

Equally ancient is the hiding of snakes beneath flowers, a fact of nature of which primitive humanity, shoeless and un-clad, had, no doubt, frequent sad experience.   The "serpent in red roses hissing," is, therefore, ubiquitous in verse.

> " Passions among pure thoughts hid
>   Like serpents under flowerets sleeping."

> " Unwholesome weeds take root with precious flowers,
>   The adder hisses where the sweet birds sing."

> " Oh, in thy truth secure, thy virtue bold,
>   Beware the poison in the cup of gold,
>   The asp among the flowers."

> " Distrust
> The vain pretence ; the smiles that harbour grief
> As lurks the serpent deep in flow'rs enwreath'd."

> "See ! how vain pleasures sting the lips they kiss,
> How asps are hid beneath the bowers of bliss."

Thence the idea of treason "Gaunt as a serpent : "—

> " Right as a serpent hideth him under floures,
>   'Til he may see his time for to bite,"

says Chaucer, or, as Shenstone puts the same thought :—

> " My tend'rest glances but the precious flow'rs
>   That shade the viper while she plots her wound. "

And so (as in Churchill) to treason particular :—

> " Those vipers to their king,
> Who smooth their looks and flatter whilst they sing."

> " Thus treason works ere traitors be espied,
>   Who sees the lurking serpent."

In Shelley the hurricane and lightning are both snakes,
and in several poets, flame—coiling, wreathing, darting,
hissing, fork-tongued—is a snake.

> " The hurricane came from the west and passed on
>   By the path of the gate of the eastern sun,
>   Transversely dividing the stream of the storm ;
>   As an arrowy serpent pursuing the form
>   Of an elephant, bursts through the brakes of the waste."

In Darwin the great sea-worm [1] is blue :—

---

[1] The sea-snake is fully referred to in later pages.—P. R.

> " Two serpent-forms incumbent on the main,
>   Lashing the white waves, with redundant train,
>   Arch'd their blue necks, and shook their tow'ring crests
>   And plough'd their foamy way with speckled breasts ;
>   Then, darting fierce amid the affrighted throngs,
>   Rolled their red eyes, and shot their forked tongues."

This cerulean snake is a recurrent figure. Science knows more than one " blue " snake, though they are not all really such. Who, for instance, that has ever seen one, would think of calling the " Korait " " *cæruleus* " ? Science, however, calls the dull, lead-coloured crow of India " *splendens.* " So let it pass. The poets, however, have " blue " snakes which they mean to be really blue. Many have water-snakes of this colour—for Virgil had such. Darwin, Mary Howitt, and Shelley have, with a natural license, blue sea-snakes. Others have land reptiles of the same colour. Thus in Heber's admirable rendering of Pindar's address to Agesias of Syracuse :—

> " Two scaly snakes of azure hue
>   Watched o'er his helpless infancy ;
>   And, rifled from the mountain bee,
>   Bare on their forky tongues a harmless honey-dew."

King gives Megæra the Fury a ringlet of blue snakes, and in Congreve the Gorgon's headdress of vipers is " blue as the vault."

While on this subject, it is very curious that the poets should perpetually speak of the " coronal " of snakes—whatever they may mean by it. Yet there is a whole genus scientifically named " coronellinæ."

CHAPTER VI.

"THE TUNEFUL FROG."

> " From the hay-cock's moistened heaps
> Startled frogs take vaunting leaps ;
> And along the drewen mead,
> Jumping travellers they proceed ;
> Quick the dewy grass divides,
> Moistening sweet their speckled sides ;
> From the grass or flowret's cup
> Quick the dew-drop bounces up."—*Clare.*

" A LONG line is run to make a frog "—by which Sir Thomas Browne pleasantly expresses his admiration of the series of metamorphoses by which the frog arrives at complete individuality.   First of all, that " lentous and transparent body," full of " little conglobulations," which we call spawn :—

> " Ere yet with wavy tail the tadpole swims,
> Breathes with new lungs, or tries his nascent limbs,
> Her countless shoals the amphibious frog forsakes,
> And living islands float upon the lakes."

And then the " porwigle " or " tadpole," all tail and head. By-and-by the thing sprouts two hind legs, still keeping its long caudal appendage.

> " So still the tadpole cleaves the watery vale
> With balanc'd fin and undulating tail."

Next it grows its fore-legs, and swims about as a long-tailed froglet.   Then we see it sitting on the bank with only a short stumpy tail.   Return two days later, and, lo ! the tail has gone altogether, and a tiny " frog " is there.

> " New lungs and limbs proclaim his second birth,
> Breathe the dry air, and leap upon the earth."

"Frogs and toads and all the tadpole train" are un-popular with the poets. They dislike their appearance and detest their voice. They remember, too, against them the description in Holy Writ of "the croaking nuisance" of Egyptian chastisement :—

> " The river yet gave one instruction more,
> And from the rotting fish and uncococted gore,
> Which was but water just before,
> A loathsome host was quickly made
> That scal'd the banks, and with loud noise
> Did all the countryside invade,
> As Nilus when he quits his sacred bed
> (But like a friend he visits all the lands)
> With welcome presents in his hands.
> So did this living tide the fields o'erspread.
> In vain th' alarmed country cries
> To kill their noisome enemies ;
> From th' unexhausted source still new recruits arise.
> Nor does the earth these greedy troops suffice ;
> The towns and houses they possess,
> The temples and the palaces,
> Nor Pharaoh nor his gods they fear,
> Both their importune croakings hear ;
> Unsatiate yet, they mount up high'r,
> Where never sun-born frog durst to aspire,
> And in the silken beds their slimy members place,
> A luxury unknown before to all the watr'y race."—*Cowley*.

In the New Testament the batrachian folk are only once mentioned. "And I saw," says St. John in the Revelation, "three unclean spirits like frogs come out of the mouth of the dragon and out of the mouth of the Beast, and out of the mouth of the False Prophet." In the Old they recur three times, and always in the same association, as the instrument of Osirtesen's humbling :—

> " That croaked the Jews
> From Pharaoh's brick-kilns loose "—

on the day when his borders were smitten with frogs—

"that loathed invasion," as Milton calls it—which "the river brought forth abundantly."

" A race obscene
Spawned in the muddy beds of Nile,
Polluting Egypt.   Gardens, fields, and plains
Were covered with the pest.   The trees were filled,
The croaking nuisance lurked in every nook ;
Nor palaces, nor even chambers 'scaped ;
And the land stank, so numerous was the fry."—*Cowper.*

With this point of high prescription, sanctified by a supreme authority, the poets are accurately well content, and the frog remains therefore "loathsome."   "Puffing" frog, "dew-sipping," "sly-jumping," are found among the more curious epithets applied to the animal.   But "speckled," "hoarse," and "slimy" are the more common.

Yet if more poets had been familiar with zoology they might have fairly revelled in the beauties and significances of the frog-world.   Of the strange chain of metamorphoses by which the "porwigle" laboriously graduates in maturity I have already spoken, but what shall we say of the Pipa that hatches her young out of dimples upon her back? She has, as it were, a false skin, and under this are little pits, in each of which an egg germinates.   By-and-by they hatch.   The young ones creep out through the upper skin on to the mother's back, and hop off to the ground.   She then casts her old skin and starts afresh.   Or of the Alyatis, where the husbands share with their wives in a fair and manly way the inconveniences of reproduction, and "lie in," so to speak, of one half the eggs while the mother takes charge of the other half?   Or of that Hyla whose eggs the male takes up in its paws and packs away comfortably into a pouch on the mother's back, where they hatch?

Were the responsibilities of parents ever more conscientiously undertaken?   Then, too, the originality of such proceedings !   The poor things have got neither "dens" nor "nests," and they refuse to leave their eggs lying about, as some things do, at the mercy of the spoiler. So they always carry them with them, the father dividing the work with the mother. This is surely admirable. Yet, as Professor Martin Duncan says, "Frogs have little to thank humanity for." Children tease them, ignorant adults persecute them. Men of science delight in microscopical and galvanic experiments with them. Birds and beasts and fish and snakes are perpetually hunting them. In the water and out of it they are for ever under pursuit. Nor does much sympathy attach to them in their afflictions—because they are only frogs.

Yet if regard attaches, as it certainly should do, to that which is eminently edible, we owe some consideration to this creature. It is particularly good eating, and by the draining and cultivation of the country, we have lost a side-dish which the Continent and America more judiciously appreciate. Again, if respect is due to that which is useful, the utmost deference should be shown to frogs and toads, who are the most relentless enemies of insects injurious to plants, and a terror to all evil-doers in gardens and orchards. Is admiration the prerogative of beauty? Then where can it be better bestowed than on the pretty green and red frogs of South Africa, or the cornfield frog of Carolina, which is dove-colour above and silver below, or the exquisite tree-frogs, grass-green and gold as a rule, but in Central America sky-blue above and rose-pink beneath? They are all living gems, and science calls them by pretty names—aurea, cœruleus, ornatus, pulcher, and elegans. Even in mere variety there is a pleasant virtue : for this we load the variable chrysanthemum with compliments. Yet among the frogs and toads there is a diversity almost as bold and quaint as amongst the ochids, while for positive beauty they are, some of them, unsurpassed.

As for their voice, the poets have much excuse for disapproval.

"Gossips are frogs—they drink and talk."

I have myself wondered that any one could ever have spoken
with admiration of their song, and am not surprised that
the Abderites should have been driven from their homes
by the intolerable monotony of the batrachian chorus.
How it exasperated Bacchus on his way to Hades—that

> " Brekekex coax coax, brekekekex coax coax,"

of the persistent multitude !

In Menu's " After-world " there are twenty-one purgatories.
One of them is filled with mud; and if the mud be filled
with frogs, I think I would rather be consigned to any one
of the other twenty; albeit, I know that Indra's august
abode is enlivened, according to Hindoo legend, by "the
harmonious voices of the black bee and the frog."   So, too,
in Aristophanes, Charon, laughing, says—

> "You shall hear most delightful melodies as soon as you lay-to at your
>   oars.
> From whom?
> From swans—the frogs—wondrous ones."

And the frogs have much to say in their own praise :—

"Marshy offspring of the fountains we, let us raise our
voices in harmonious hymn —*brekekekex*—in sweet-sounding
song—*coax coax*.   Thus sung we in the marshes by the
Acropolis, making festal the rites of Nisæan Bacchus.
Brekekekex coax coax.   The Muses of the beautiful lyre love
us—*coax coax*—and so does horn-footed Pan who pipes
upon the reed—*brekekekex*—and Apollo, the sweet harper
—*brekekekex coax coax.*   So let us sing and leap, and leap
and sing again, through galingal and sedge, chanting as we
dive our choral strains to the music of breaking bubbles—
brekekekex coax coax."

Other poets, however, are not of Pan's opinion, nor of
Apollo's.   They recognise no harmony in the voice of the
batrachians.   Southey quotes it as the extreme antithesis of
melody.   Spenser, in his " Epithalamium," warns them off—

> " Ne let th' unpleasant quyre of frogs still croking
> Make us to wish theyr choking."

"Dutch nightingales" is a popular nickname of these loquacious amphibians, and Allan Ramsay derisively rallies the Hollanders upon their songsters.

Elegant Paris, however, has a better claim to these mud-larks, as I may call them. For three frogs once formed the civic device of Lutetia—"the mud-land "—

> " Where stagnant pools and quaking bogs
> Swarmed, croaked, and crawled with hordes of frogs ; "

but in Clovis' time the grenouille was "miraculously " transfigured into the fleur-de-lis, one product of the marshes thus supplanting the other upon the banner and shield of France. The truth, perhaps, is that about that time our neighbours discovered what excellent eating their national device was, and not caring to emblazon that which they cooked, they promoted the frog from their oriflamme to their stew-pans. The Moon-folk, however, had anticipated them, for, so Lucian avers, "they used but one kind of food." "There are," he says, "great multitudes of frogs flying about in the air ; these they catch, and, lighting a fire, cook them upon the coals ; and while the frogs are a-cooking they sit round the fire, just as men sit round a table and swallow the smoke, thinking it indeed to be the finest thing in the world."

"Soulless " is a good epithet (of Mackay's) for the croak of the creature, as any one who has listened long to their unmeaning clamour will confess, but I like Moore's humorous rendering of its significance none the less :—

> " Those frogs whose legs a barbarous cook
> Cut off, and left the frogs in the brook
> To cry all night, till life's last dregs,
> ' Give us our legs—Give us our legs.' "

Any translation of the sound that makes sense of it commends itself to me. For I should be glad to be convinced that the moist and garrulous things had souls. They have got calves to their legs, a feature which, if I am not wrong, no other animal but man possesses.

Yet, when in company, they have a wide range of expression from the crisp, shrill chirrup of the tree-frogs, to the loud snore of the "Cambridge nightingales." The multitudes of the Arkansas swamps have a nasal metallic "yank-yank," as different as possible from the deep "owk-owk" of the French frogs. The fire-bellied toad has a clear, resonant voice, the bull-frog a profoundly sonorous one. The natter-jack cries "gloo-gloo," the green toad "may-may," while, for the noises of the rest, the frog-chorus of Aristophanes, already quoted, renders them faithfully enough by brekekekex and coax coax.

Again and again in legends they are struck dumb; now by saints for disturbing devotions, and now by nymphs for defiling fountains. On the other hand, men are punished by being made to croak—

> " As when those hinds that were transformed to frogs
> Railed at Latona's twin-born progeny."

In Syriapha, blessed isle, the frogs were voiceless; and it is a fact that the illustrious Mecænas bore a frog for his device *as an emblem of taciturnity*, borrowed from those batrachians on the shores of the Ægean, who " never croak in their own marshes," according to Ælian. " Mute the same is, and never croaketh," saith Pliny. Emblematic, therefore, of silence and secrecy, qualites for which Augustus held Mecænas in such respect and favour.

The " bull-frog," which I take to be a corruption of bell-frog, let philologists say what they like, is not an aversion of the poets. It is "deep-mouthed and doubly harsh" in Byron, and Faber has—

> " Beneath thy feet
> A lonely bell-frog from the reedy fosse
> Rang his distinct and melancholy fall,
> But harshly to thy travel-wearied mind
> Most soothingly attuned."

There ought to be nothing laughable about the creature's voice, but I confess I have laughed consumedly at its pompous gravity; and a friend once told me how a love-making scene, which both the lady concerned and he meant to be most serious, was made utterly ridiculous, in consequence of a bull-frog chiming in most inopportunely, whenever sentiment demanded silence, or, at most, only a nightingale's song. However, the frog was, after all, a true friend, for the marriage, which eventually followed their laughing betrothal, has been a happy one, owing, so they say, to "that old bull-frog" having stopped them both from committing themselves and each other to "a lot of bosh" at the commencement.

In fairy-stories the frog is perpetually recurring. Its shape is popularly held in aversion, so nothing could be more suitable for the utmost degradation of enchanted princesses and princes. Ivan the Tsar's son has to marry a frog who eventuates blissfully in a very Helen of personal charms. In Grimm, the exquisite princess has to wed a frog which turns into an adorable prince. When the lovely maiden is to be transformed into an odious object, toads fall from her mouth when she speaks. Yet in folk-lore, frogs are uniformly beneficent. One brings the Queen her Rose-briar daughter; another, a fat old frog, makes Dummling's fortune for him.

So, too, are the toads. They are always bringing good luck to children, or treasures (as "the toad with the crooked leg" did) to princesses. As the metamorphoses of human beings, they are intended to be repulsive. In their own persons they are benign. As the familiars of witches the paddocks had a bad name—

> " I went to the toad that breeds under the wall ;
> I charmed him out, he came to my call.
> I scratched out the eyes of the owl before ;
> I tore the bat's wing—what would you have more?"

But, like Robin Goodfellow, who was also the companion of hobgoblins and all manner of Serene Naughtinesses, they exercised their power with consideration and benevolence.

Wordsworth regards the frog as a sort of amphibian Mark Tapley, and sees in the creature jumping about on a wet day a moral of cheerfulness under depressing circumstances, and bids his readers

> " Learn from him to find a reason
> For a light heart in a dull season."

But considering that Wordsworth professed an exceptional sympathy with nature, it is curious that he should have missed sense by such a distance.   Wet days are, of course, the frog's gayest weather ; then it picnics, flirts, puffs out, is happiest.   As a matter of fact, Wordsworth knew this very well.

> "On shore the coming deluge draws the race
> Of reptiles from their haunts, in mead and grove,
> Concealed—the puffing frog, the horned snail,
> And all the species of the slimy tribes."—*Grahame.*

Indeed, in myth, they are often clouds and pluvial.   Their croaking is the rain signal.   They are "rain-desiring," so "rain-heralding," and thus come to be "rain-compelling."

Of the actual frogs of story the poets have three groups. There are first those "good Æsop's frogs" that asked for a king—

> "Thus kings were first invented, and thus kings
> Were burnished into heroes, and became
> The arbiters of this terraqueous swamp,
> Storks among frogs."

—and afterwards changed the dynasty—

> " Loud thunder to its bottom shook the bog,
>   And the hoarse nation croaked 'God save King Log.'"

From the same delightful source comes that other that would have her children know that their mother could make herself "so big as anything"—

> " And thus the reptile sits
>   Enlarging till she splits."

And also those, the "much-complaining" frogs, that presumptuously entertained the idea of punishing the Sun for drying up their mud.

Next, in frequency of mention, are the pugnacious frogs of the classics that did battle with the mice, where "he who inflates the cheek" warred with the "cheese-nibbler;" and those others who went, as Beattie sings, to battle against the cranes :—

> " And there the frog, a scene full sad to see,
>   Shorn of one leg, slow sprawled along on three.
>   He vaults no more with vigorous hops on high,
>   But mourns in hoarsest croaks his destiny."

The third group are those of Holy Writ, to which I have already referred.

CHAPTER VII.

"THE LOATHED PADDOCK."

" Would'st thou think that toads or snakes or efts
Could e'er be beautiful?"—*Shelley.*

"INASMUCH," says De Gubernatis, "as the toad is a form
proper to the demon, it is feared and hunted; inasmuch
as, on the contrary, it is considered as a diabolical form
imposed upon a divine or princely being, it is respected
and venerated as a sacred animal." In poetry, the toad
has only one, the "loathed" aspect; although in popular
estimation in all countries it has both a sinister and a
benign.

It is a lucky omen if one crosses the path of a wedding
party bound for the church. Did not St. Patrick—all
pictures to the contrary notwithstanding—spare them when
he drummed "the vermin" out of Ireland? Just as in
Cornwall a man may not shoot a raven lest he should kill
King Arthur unawares, so in Tuscany, you may not hurt a
toad lest you should do a mischief to some young princess
or heroine who has been cruelly transformed into that shape,
and who is only waiting for the beautiful prince to come,
when the maiden will resume her charms and "live happily
ever afterwards." In the folk-tale of some countries, the
Beast who marries Beauty is a toad, and many stories
substitute this creature for the frog in such stories as where,
benign and amiable, it fetches rings up for sultans' sons

and balls for kings' daughters. "Medicinally, the paddock had once a high value for cancer, and in Europe it is still worn on the person as a charm against poisons and the plague. On occasion, too, it was a potent beast. For if it found a cock's egg and hatched it, the result was a cockatrice, a fearsome thing, which of its own accord grew a crown on its head and so became a basilisk, and could kill by merely looking. A very notable worm indeed, and most reverend, was "this crowned asp." Moreover, the toad,

> " Though ugly,
> Wears yet a precious jewel in its head."

In that very fascinating work, "The Natural History of Gems,"[1] a chapter is devoted to those "stones of virtue," which were supposed in olden times to have been produced by, or found inside of, beasts, birds, fishes, and reptiles— such as the hyæna, which, placed under the tongue, conferred the gift of prophecy, and the marvellous "lynx-stone;" the grass-green chloritis, found only in wagtails, and the alectoria, a crystal formed inside cocks; the cinædia, developed in the head of the fish so called; the draconite, dreadfully lodged in snakes, and the famous "batrachite" or "bufonite"—the "toad-stone." This last was said to be of three kinds,—one yellow and green "like a frog," the second black, the third red and black. This tradition being handed down to mediæval fancies resulted in the toad being credited with "a jewel in its head," which was variously called "borax," "nosa," and "crapodinus."

> " The unwieldy Toad
> That crawls from his secure abode
> Within the mossy garden wall
> When evening dews began to fall,

---

[1] By C. W. King, M.A., Fellow of Trinity College, Cambridge.

<blockquote>
Oh mark the beauty in his eye:
What wonders in that circle lie!
So clear, so bright, our fathers said
He wears a jewel in his head!"
</blockquote>

This rare gem was a specific against poisons, and a great number of rings are in existence in which the "Kroten-stein' is set as a talisman against venom. Erasmus writes of a famous toad-stone dedicated to our lady of Walsingham, and numerous mediæval jewels now preserved in collections owed their great value in the past to the magic potencies supposed to be vested in the mysterious "stone." It has been discovered by modern investigation that these bufonites are really the bony plates lining the jaws of some fossil fish, hemispherical bosses which served the finned creatures for teeth ; but fortunately this discovery was not made in time to spoil the pleasure which our forefathers took in their "toad-stones."

On the other hand, "in its diabolical aspect," Bufo has many depreciative associations. It was said to spit poison, and to shoot it out at its pursuers, also to envenom all the plants it passed over. This is the poets' acceptation of toads. Indeed, they are so venomous that serpents "of most deadly sort" are bracketed with them, as being—in Blair for instance—the "superlatives" of evil.

<blockquote>
" Toads and snakes, and loathly worms,
And venomous and malicious beasts, and boughs
That bear ill berries."
</blockquote>

Eliza Cook calls them "foam-spitting" and "vile," and this unfortunate lady had, at any rate, such privilege as might be begged from the precedent of Milton, living two centuries before her, saying that Satan

<blockquote>
" Squat like a toad close to the ear of Eve."
</blockquote>

Coleridge, always punctual in plagiarism, has "Slander squatting near, spitting cold venom in a dead man's ear," altogether a ridiculous image and a waste of "venom." In Dyer we find them associated with several poetical horrors —a curious assortment.

> " 'Tis now the raven's dark abode,
> 'Tis now th' apartment of the toad,
> And there the fox securely feeds,
> And there the pois'nous adder breeds."

Churchill is in the same vein —

> " Marking her noisome road
> With poison's trail, here crawl'd the bloated toad ;
> There webs were spread of more than common size,
> And half-starv'd spiders prey'd on half-starv'd flies ;
> In quest of food efts strove in vain to crawl ;
> Slugs, pinched with hunger, smear'd the slimy wall ;
> The cave around with hissing serpents hung ;
> On the damp roof unhealthy vapours hung,
> And famine, by her children always known,
> As proud as poor, here fixed her native throne."

Moore is characteristically fanciful—

> " There let every noxious thing
> Trail its filth and fix its sting ;
> Let the bull-toad taint him over,
> Round him let mosquitoes hover,
> In his ears and eye-balls tingling,
> With his blood their poison mingling."

This "bull-toad" is thoroughly Moore ish, and belongs of course to the same poetical family as the "night-raven" or "wood-wolf." Toad and adder, by the way, is a very frequent association—showing how thoroughly the fiction of the poisonous character of the toad had taken hold of the poets' fancy. So, too, had the wickedness and the blood-thirstiness of the owl.

"Full-blown" is Pope's delightful epithet for Bufo, "puffed by every quill." It is not what he meant, of course, for that was "inflated," "puffed-out," "bloated." But it conveys his meaning admirably none the less, and has no spite in it. "Fulsome" is one of Dryden's epithets, and, even more absurdly than "full-blown," suggests an excessive distension. The real meaning of the word, however, is "nauseating," "nauseous," and differs, therefore, from Pope's in being extremely rude to Bufo. "Slow, soft toad," says Shelley—an excellent phrase. But the majority, from Spenser to Wordsworth, have only "bloated" and "loathly." Moore calls it "obscene;" Southey "foul;" Savage "loathsome," and so on. Thomson, of course, is egregious as usual in infelicitous description.

In metaphor the toad comes off poorly. Spenser sets the fashion :—

> "Envy rode
> Upon a ravenous wolf and still did chaw
> Between his cankered teeth a venomous tode
> That all the poison ran about his maw."

In Lovelace's duel :—

> "First from his den rolls forth that load
> Of spite and hate, the speckled toad,
> And from his chaps a foam doth spawn."

Pope, in his prologue to the Satires, has Sporus, who, a

> "Familiar toad,
> Half froth, half venom, spits himself abroad."

Thomson in his "Castle of Indolence" has "a wretch :"—

> "A wretch, who had not crept abroad
> For forty years, no face of mortal seen,
> In chamber brooding like a loathly toad."

Southey in "The Miser's Mansion" makes the toad a feature of its horrors :—

> "Thy tall towers tremble to the touch of time,
> The rank weed rustle in thy spacious courts;
> Filled are thy wide canals with loathly slime,
> Where, battening undisturb'd, the foul toad sports."

As opposed to all this fancy, how strangely reads this

extract from Sir Joseph Banks, not the only naturalist by a
score who has left on record his admiration of and liking
for toads :—

"In a cage, if properly provided with a damp corner,
lined with mat, and kept clean, it will live happily and comfortably a long time.   Feed it once a day with earthworms,
maggots, or flies ; of all of which it is very fond.   I conceive
that the workings of your reason would soon gain a victory
over your prejudice if you could conquer your first disgust
and look at the animal with any kind of indifference.   The
eye of the toad would first attract your attention, which
is brilliant and intelligent; his action in seizing his prey,
which is composed of a mixture of force and cunning, would
amuse you, and I am confident, if the first prejudice were
removed, that in one week's time a toad would become an
object of amusement instead of disgust."

Its nominal connection with the toadstool is very curious,
for it arises from the word " pogge," which means a toad,
and is also an equivalent for Puck.   Now, the " pogge," as
I have already said, has in popular superstition a diabolical
aspect, and is of the nature of devilkins, of whom Puck is
the chief, and the fungus in question, by a very allowable fancy, is called indifferently a " puck's stool," or
toad-stool, and the puff-ball is a pixie-stool, or paddock-
stool.   Spenser's couplet on this connection, is surely
delightful—

> " The grisly toadstool grown there might I see,
> And loathèd paddocks lording on the same."

" Toad-flax," one of the very prettiest of all our wild-
flowers, and " toad-pipe," are said to be only accidentally
associated with the toad, as the original spelling was " tod,"
which means a bunch.   Not that I care in the least for such
learned enlightenments.   They spoil a great deal of pleasant
fancy, I think, by taking away its flax and its pipe from the

toad. Why should not the quiet-living, home-staying thing spin flax or use the hollow tubes of the equisetum (as the frogs pretend they use the reeds) for mellowing their utterances? There are toads that croak as melodiously as frogs.

THE POETS' FISHES.

PART II.

THE POETS' FISHES.

CHAPTER I.

FISH-MONSTERS AND MYTHS.

IT is not easy with poets' animals to draw the line between
fish and flesh. Their water creatures are so finely graduated,
from the complete human to the complete fish, that it is
difficult to say where the actual severance of species occurs.
Out on the sea-wave yonder a water-sprite "sits dodging
the blessed bird," and here on the river-bank the brown
kelpie lies, "lurking 'mid the unrustling reeds." But you
cannot see whether they have legs or tails. Yet they are
the kindred of the mermaids who wear scales on half their
bodies, and the mermaids again are the relatives of creatures
that are entirely fish.

Sometimes these sea things are dreadful, living as they
do in an ocean "scourged by hissing gales, and writhing in
their glistening coats of clashing scales," and "haunting
those dark rocks which, when the storm is up, bellow and
gnash and snarl together." Yet as a rule the poets make fun
of them. Proteus and his sea-calves have little dignity.
Triton with his puffed-out cheeks is usually laughed at.
In a general way, of course, there is the utmost, reverence
expressed for "the huge sea monsters that lift the deep
upon their backs;" and now and again there is special awe

of some particularly grim creature such as Spenser's "griesly wasserman," that

> " Makes his game
> The flying ships ; "

or " the horrible sea satyre " " that doth show his fearful face in time of greatest storm ; " or " the greedy rosmarines with visages deforme." Land monsters are all well enough. Amphibious ones just pass muster. But purely aquatic monsters are absurd. The poets, it is true, affect to believe in the fiction that the water holds an exact counterpart of the geology of the dry land—

> " For seas as well as skies have sun, moon, stars ;
> As well as air—swallows, rooks, and stares ;
> As well as earth—vines, roses, rushes, melons,
> Mushrooms, pinks, gilliflowers, and many millions
> Of other plants, more rare, more strange than these,
> As very fishes, living in the seas ;
> As also rams, calves, horses, hares, and hogs,
> Wolves, urchins, lions, elephants, and dogs,
> Yea, men and maids, and, which I most admire,
> The mitred Bishop and the cowlèd Friar,
> Of which examples, but a few years since,
> Was shown in Norway the Polonian prince."

' But they treat the marine Fauna of Fancy individually as inferior imitations of their terrestrial equivalents.

It was a common myth this that the sea held a duplicate of every animal on the earth, and antiquity therefore was familiar with many marine equivalents for their land beasts, even though they could find no better resemblances for the corresponding terrestrial beasts than a lobster for the " lion," a crab for the " bear," a skate for the " ox," a dog-fish for the " dog," and an eel for the " wolf." The names were probably given at first simply to indicate a single point of fancied resemblance, but eventually some imaginative theorist, seeing so many correspondences recognised, hit upon the idea of extending the identities throughout creation. The attempt, however, was a complete failure, and the further inquiry is made, the wider become the differences between the inhabitants of the water and the earth.

> "Where, like our moderns so profound,
>   Engag'd in dark dispute,
> The cuttles cast their ink around
>   To puzzle the dispute.
>
> Where sharks like shrewd directors thrive,
>   Like lawyers rob at will,
> Where flying fish, like trimmers live,
>   Like soldiers, swordfish kill.
>
> Where on the less the greater feed,
>   The tyrants of an hour,
> Till the huge royal whale succeed
>   And all at once devour.
>
> Thus, in the moral world we now
>   Too truly understand
> Each monster of the sea below
>   Is match'd by one at land."—*Ch. Pitt.*

Sailors and fishermen still retain many of the old names, and popular usage has familiarised us more or less with the sea-horse—the quaint little creature, so like a knight on a chess-board—sea-lion, sea-bear, sea-cat, sea-eagle, sea-bat, sea-hedgehog, sea-leopard, sea-mouse, sea-scorpion, sea-snipe, sea-swallow, sea-parrot, and so forth, while heralds are responsible for the perpetuation of many amphibious hybrids. But this tendency to see in the water a reflection of everything on land is only an instance of human self-consciousness, for if we were to be just to our seniors in creation, and more modest, we should call ourselves land-manatees, our elephants land-whales, and our tigers land-sharks. As Sir Thomas Browne says—"If we concede that the animals of one element might bear the names of those in the other, the watery productions should have the pre-nomination."

The poets' sea-horses are veritable things of saddle and harness, that "snort amayne, and from their nostrils blow the brynie stream." Again in Darwin "Her playful sea-horse wooes her soft command, Turns his quick ears, his webbèd claws expands."

These are obviously the actual chargers that bore the sea-god through the tumult of the Ægean, with such "a rushing voice of waters" as silenced for a while even the thunders of an embattled Olympus.

The sea-dog, again, is evidently no barking thing like the horrible familiars of Scylla, for the hounds of the Sicilian ogress are always mentioned specifically by name. It is a horrible fish of the shark kind—a dog-fish. Thus Moore has the sea-dog "that doats upon the small sweet fry that round him floats," and "that tracks the bloody way" of the bark. In Faber they are the companions of the seal—

> " And on the sandbanks of the bay
> Sea-dogs and seals together lay,
> As though the hot mist of noon were sweet
>   After the day's cold gloom.
> They slept like the dogs at the marble feet
>   Of a Templar on his tomb."

And from this we pass to the "sea-calf," which, in Savage for instance, basks on the rock where seals congregate for sunshine, and to the "sea-lion" that roams in Rogers "in coral groves silent and dark," and so to the seals themselves.

Poets oddly enough detest the seal. Keats only has it sadly "in death, on the cold ice with piteous bark, lie full of darts." But with the rest it is the "stinking" seal in Spenser and others, and the "misshapen" seal.

Sometimes we read " with seals and orcs in sunless caverns bred ;" at others the poets call them "orcs" when they wish to aggravate their "obscenity ;" and Leyden accuses them of attacking children by throwing stones at them :—

> " Nor ravenous seal, that suckleth on the shore
> Her hairy young, unaw'd by eye of man ;

> Her snorting oft, at sunset, on the coast
> Of Argus, fruitful land of vital grain,
> The wanton damsel mocks, and children join
> Insultant, to provoke with rustic names;
> Suddenly awak'd she starts with uncouth gait
> Gleaning their steps, and now with either hand
> Gathers, and throws full fast stone after stone;
> Blind with revenge, nor mindful of her end,
> Though near; for now her neck the hissing ball
> Hath pierced. . . .
> In vain, alas! thou homeward hasteth back,
> Mad to have risked thy life with artful man
> On land, to thee strange element, where now
> Thou agonising liest, a monument
> For others not to dare their sphere beyond;
> While children, from their flight retiring, gaze
> And wonder at the shaggy monster's form."

The seal is the "dire" Phoca of poetry, sometimes a "scaly" beast that (as in Spenser), draws the shell chariots of sea divinities, but more often an ordinary seal, found in the high seas, rolling in company with whales and "grampi,"—as more than one poet calls them.

"Old sea-lion of the North" is an excellent conceit of Mackay, speaking of the Viking.

* * *

It is characteristic of poets to prefer for their purpose those "fish" that are associated with old myths, and have classical associations, rather than those that are simply picturesque in nature. The dolphin, nautilus, and Behemoth are therefore especial favourites, while as terms of reproach the shark and pike are of frequent occurrence. Yet all the same, when we have allowed for the normal prepossession of bards in favour of the ancient and mythical, heraldic or absurd, there remains a considerable balance of very good description of, and pretty thought, about the real water-folk.

In the general struggle some of them have attained to

honours by their force of character. For instance, the salmon, so lordly in its nature as to worthily justify the name of that proud King of Elis who defied Olympus. But he was hurled to the shades by a judiciously directed thunder-bolt, and thus abundantly expiated his arrogant obliquities. So too the shark, that awful Attila of the sea ; and the pike also, the " dispeopler of the lake," that by its ferocity of countenance and manners usurps the autocracy of the reedy pond. Other fish again have compassed dignity by the passive virtues of their flesh. Did not Domitian order a special session of the Senate to discuss the cooking of a turbot, and "nihil ad rhombum "—all Lombard Street to a China orange—pass into a proverb ? What man in Rome would not have been a lamprey to be petted by the beautiful wife of Drusus ? and what a pitch of dignity they attained to in the households of epicures, those mullet, and muræna, and carp. But by far the greatest number have achieved distinction by legendary exploits, or by accidents of honour. Thus the dolphin and the tortoise, or the haddock and the John Dory. It was a crab that retrieved the crucifix of St. Xavier from the sea—

> " Nor let Xavier's great wonders pass concealed,
> How storms were by the Almighty wafer quelled,
> How zealous crab the sacred image bore,
> And swam a Catholic to the distant shore "—

and to a codfish that Scandinavia owed its recovered crown. Was it not a fish that guided the Vedic ark to its resting-place, the hill-peak Naubandha ?—

> " In the whole world of creation,
> None were seen but these seven sages, Manu and the Fish.
> Years on years, and still unwearied, drew this Fish the bark along,
> Till at length it came where reared Himavan its loftiest peak ;
> There at length they came, and smiling, thus the Fish addressed the Sage :
> ' Bind now thy stately vessel to the peak Himavan ! '
> At the Fish's mandate, quickly to the peak of Himavan
> Bound the Sage his bark ; and even to this day that loftiest peak
> Bears the name Naubandha,"

and from a fish-pond (according to Arabic legend) that Moses was rescued by Pharaoh's daughter? When the demons had usurped Solomon's throne, and the monarch was an outcast in his dominions and jeered at as a sort of Perkin Warbeck, a preposterous claimant, a fish found the omnipotent signet-ring, and so enabled the king to reascend his throne. Did they not give their names to a score of cities? Is not fish one of the special foods promised to the faithful in the paradise of the Moslem, with, hard by, that tree from Sinai that yields sauces "appropriate thereto, for them who eat"—a kind of paradisiacal cruets.

In character they range through every variety of temperament, from the gentle carp that in Java and elsewhere are tamed into the playfulness and familiarity of dormice or caged birds, or the Adonis, "darling of the sea,"[1] to the dog-fish, that are cruel and fierce beyond all mammalian comparison. And what episode is there in all human knowledge more terrible than the manner of the death of those whales which the dog-fish follow for days and days and days, living upon them as they go?

Fish are, as a rule, in myths, the stupid persons of the narrative, the foolish folk who are found dancing in the nets just when they should be most serious; who get caught and beg the fishermen to put them back, "so that we may grow larger and better worth your eating;" who catch hold of the hook in order to pull the angler into the water; who rush into the net just to make fun of the fishermen,

---

[1] "There is also," saith the "Compleat Angler," "a fish called by Aliaut the Adonis, or Darling of the Sea, so called because it is a loving and innocent fish, a fish that hurts nothing that hath life, and is at peace with all the numerous inhabitants of that vast watery element; and truly I think most anglers are so disposed to most of mankind." Wait for the adonis that Marc Antony and Cleopatra used to make those notable fishing-excursions, at once the admiration and the scandal of "surrounding nations."

forgetting that, though it is the same old net with the same meshes they used to slip through when they were tiny fry, they have been gradually getting bigger themselves,—who fall victims, in fact, to every designing person that comes their way.

CHAPTER II.

FISHES OF THE ANGLE.

POETS are in sympathy with anglers, whether, as children,
they go with Clare—

> " Chasing the stickle o'er the shallow tide,
>   And flat stones turning, where the gudgeon hide "—

or, in completer accomplishment of adult sport—

> " Where the duck dabbles 'mid the rustling sedge,
>   And feeding pike starts from the water's edge "—

with Mackay—

> " Rod in hand they go
> To streams that leap too frolicsome to flow,
> Angling for trout, and catch them by themselves,
> In fancied citadel, beneath the shelves of slippery stone ; "—

all of which is very curious.   They do not, it is true, like
to see men putting the poor worms on to their hooks, but,
for poets, they are singularly insensible to any suffering that
may be caused to the fish.

> " When, if or chance or hunger's pow'rful sway
>   Directs the roving trout this fatal way,
>   He greedily sucks in the twining bait,
>   And tugs and nibbles the fallacious meat.
>   Now, happy fisherman !   Now twitch the line ;
>   How thy rod bends ! behold, the prize is thine.
>   Cast on the bank, he dies with gasping pains,
>   And trickling blood his silver mail distains.

> You must not ev'ry worm promiscuous use ;
> Judgment will tell the proper bait to choose.
> The worm that draws a long immod'rate size
> The trout abhors, and the rank morsel flies.
> And if too small, the naked fraud's in sight,
> And fear forbids, while hunger does invite.
> Those baits will best reward the fisher's pains,
> Whose polish'd tails a shining yellow stains.
>
> Cleanse them from filth, to give a tempting gloss,
> Cherish the sullied reptile race with moss ;
> Amid the verdant bed they twine, they toil,
> And from their bodies wipe their native soil."—*Gay.*

Or again, the salmon-fisher—

> " Till, tired at last, despoil'd of all his strength,
> The game athwart the stream unfolds his length.
> He now, with pleasure, views the gasping prize
> Gnash his sharp teeth, and roll his bloodshot eyes ;
> Then draws him to the shore with artful care,
> And lifts his nostrils in the sick'ning air.
> Upon the burden'd stream he floating lies,
> Stretches his quiv'ring fins, and gasping, dies."

With all their strong objections to birds catching insects, animals eating each other, and human beings hunting, shooting, and snaring game, poets speak tenderly of fishing, and admiringly of success with rod, net, or spear.

Collectively in single poems, and individually in hundreds of scattered verses, the denizens of British waters find such frequent reference as would surprise any literary angler who cared to rummage our classics for piscatorial allusions. Indeed, it would be well worth the while of some literary fisherman of the " Red Spinner " type to run over the British poets for the poetry of fishes. The mistake hitherto has been, it seems to me, to collate only "the Poetry of the Angle." Now, the majority of our piscatory bards were better anglers than poets, and their descriptions of fresh-water fishing and fresh-water fishermen are vapid, sententious, and conceited. I know nothing more depressing in the range of poetry than their odes on angling and so forth, written apparently under the mocking burden of empty creels. Izaak Walton himself is very dreary reading indeed, unless (I hope I am not profane) for one in the humour to be amused by tediously-sustained affectation. He then becomes undoubtedly "quaint." Most poets, moreover, when they go a-fishing, seem to me to become at the outset apologists for the sport by pointing out how suitable angling is for the observation of Nature—as if one could not worship the divinities of river-bank and meadow, of sedge and overhanging wood, without having a box of worms in one's pocket. I always think of that dragon-fly that comes and sits cross-wise on the tip of the rod whenever an angling poet begins maundering about Nature, and it has been a result of my study of the bards to find that angling is emphatically the pastime of the commonplace versifier. When trivial Gay tries to work himself up into enthusiasm over a salmon, he firks up such a very moderate degree of fervour as reminds one of an organist at an instrument with leaky bellows, that gives forth only "the exiguous residual squeak." Somerville was of course a true poet of the chase, but he is sportsman far before he is poet, and his descriptions are at most times of revolting cruelty. But the majority of the "poets of the Angle" will be found to be invertebrates. They hope by referring to the sky, and the grass, and the water to magnify the sport obliquely, and with a great show of commonplace observation and some piety to enhance indirectly the virtues of an occupation which of itself did not suffice to float their lines.

Such collections of verse, therefore, represent the poetry of fishing neither adequately nor becomingly. But if a tasteful compilation of the poetry of *fishes*, "from the oyle-steeped anchovie" to the "cumbrous grampus," from "the sudden rushing of the minnow shoal," scared from the shallows by a passing tread, "to those living islands, whales," were to be made, the result would be found abounding in vigorous descriptions of water life—Leyden's, for instance—without any references of any kind to anglers and most charming vignettes straight from nature, quite irrespective of rods or "worms." Such a work would be welcome not merely to sententious anglers, but to all lovers of nature, and that means virtually everybody.

My own concern just now is with fishes, and not fishermen; but the attitude of the poets towards the craft may find a passing notice, for it is rather a curious one, inasmuch as they do not seem to recognise any cruelty in the sport, but, on the contrary, to regard the catching of fish as innocent and commendable. Poetical sympathies are, no doubt, rather unaccountable. The deer is pitied, but the fox never; hares are condoled with, but not rabbits; they exult in the killing of otters, which kill fish for food, and condone the killing of pike for sport. Indeed, the killing of fish generally is amiably condoned. Sometimes there is an affectation of compromise by throwing in a word of pity for the worm, as Izaak Walton does himself—"damned neuters" (as Dryden has it), "in their double way of steering, neither fish nor flesh, nor good red herring"—and very often they use some such phrase as this—

> " With slender line
> And yielding rod, *solicit to the shore*
> The eager trout "—

to temper the actual fact; but, as a rule, there is nothing but straightforward approval of "that solitary vice;" whether only as children "catching prickly stickles in their rout, and miller's-thumbs and gudgeons driving out "—or the more serious processes of maturer angling. No poet ever reproaches children for murdering loaches, minnows, gudgeon, and similar small fry "which everywhere are found in every little beck;" although poetry is full of reproofs for doing bees, butterflies, and beetles to death.

> " The various arts how fishes might be caught,
>   Sometimes with trembling reed, and single hair,
>   And bait conceal'd, he'd for their death prepare,
>   With melancholy thoughts and downcast eyes
>   Expecting, till deceit had gain'd its prize ;
>   Sometimes, in riv'let quick, and water clear,
>   They'd meet a fate more gen'rous from his spear.
>   To basket oft he'd pliant osier turn,
>   Where they might entrance find, but not return.
>   His net, well pois'd with lead, he'd sometimes throw,
>   Encircling thus his captives all below.
>   But when he would a quick destruction make,
>   And from afar much larger booty take,
>   He'd through the stream where most descending, set
>   From side to side his strong capacious net ;
>   And then his rustic crew, with mighty poles,
>   Would drive his prey from out their oozy holes,
>   And so pursue them down the rolling flood
>   Gasping for breath, and almost chock'd with mud,
>   Till they, of farther passage quite bereft,
>   Were in the mesh, with gills entangled left."

There is nothing in these lines to hint at pitifulness in the writer, and the same want of sympathy with the fish— except on very rare occasions, where, as in Mackay, the fish are warned against their " enemy "—is characteristic of all piscatorial verse.

That the world sustained a great loss in the destruction of Solomon's work on Fishes, may be accepted as beyond dispute ; for let the scientific attainments of the sumptuous builder have been what they might, there can be no doubt of it Solomon, who was of an artistic kind, would have preserved to posterity vast quantities of old-world nonsense, possibly even of antediluvian facts, which are now hopelessly lost to us ; and except Solomon, no other personage of Holy Writ has expatiated on the subject of fishes. We have no Scriptural recognition of any great fisher " before the Lord." Indeed the untranslated Bible is singularly reticent on the subject, for it does not specify a single fish. Tobit's fish and Jonah's fish, the fishes of the Psalms and of the New Testament, are spoken of only generically, and even when the Lawgiver is enumerating the things which the Hebrews might and might not eat, he is careful to distinguish by their names the creatures in fur and feathers, but the fish are merely divided into " those with scales and fins " and " those without." Still more remarkable is it that Peter and his comrades, themselves professional fishermen, should have omitted to identify the actual species with which the Saviour worked His miracles. In fish history, therefore, there is a very considerable gap, and it is not until we go to Pagan mythology that we find the things of the water identified into species.

Poets divide them broadly, according to poetic susceptibilities, into the tyrant and the victim, the eater and the eaten; reserving always, among the latter, their tenderest sympathies for those species that are the best eating. Nothing exasperates a poet against the otter so much as that the "goose-footed prowler" should dare to eat trout. If it only ate roach, their indignation would be mitigated. If it devoured sticklebacks exclusively they would be indifferent to its poaching. When it consumes the pike they applaud it.

> " Where rages not oppression ? where, alas,
> Is innocence secure ?   Rapine and spoil
> Haunt e'en the lowest deeps ;
> Seas have their sharks,
> Rivers and ponds enclose the rav'nous pike ;
> He in his turn becomes a prey, on him
> Th' amphibious otter feasts.   Just is his fate—deserved."

But the otter has the presumption to prefer the best fish —to have, in fact, the same tastes as man himself; and so, just as they abuse the wolf for eating their mutton (they do not really sympathise a bit with the sheep), and the fox for stealing the hens that lay the eggs for their breakfast, so they, the poets, pelt the otter with every name they can think of, because it kills and eats the finer fish, salmon and trout.

A volume as large as the present one could easily be written on the Fishes of poetry alone, for it will be seen from the next few pages (which read, I fear, very much like a mere alphabetical catalogue), how comprehensive the range of poetical reference to water-folk really is. The subject has, I know, already attracted literary men of the angle, and a great deal of most excellent work has been done ; but there is not as yet any complete Thesaurus of fish-verse : than which, I can imagine, for fishermen, no more fascinating volume, nor for the curious reader outside the craft, one of more unconventional interest.

To take them in order. First, the barbel, "than whom a braver doth not swim" (Drayton), "nor (newly taken) more the curious palate please ;" the "silver broadside" bream of Quarles, that gives Waller these curious lines in the " Battle of the Summer Islands "—

> " The boat which on the first assault did go,
> Struck with a harping-iron the younger foe
> Who, when he felt his side so rudely gored,
> Loud as the sea that nourished him he roared :
> As a broad bream, to please some curious taste,
> While yet alive, in boiling water cast,
> Vexed with unwonted heat he flings about
> The scorching brass, and hurls the liquor out ;
> So with the barbèd javelin stung, he raves,
> And scourges with his tail the suffering waves "—

The "golden carp in his precious mail" (Charlotte Smith), and "yellow carp in scales bedropped with gold " (Pope), " in burnished golden mail " (Somerville) ; the chub, " which some (whose neater name) a cheven call " (Drayton), " food to the tyrant pike (more being in his power), who for their numerous store he most doth them devour ;" the " fast-feeding " cod ; the " muddy " conger ; the " pretty slender dare, of many called the dace," " with silver back "—

> "Where in the dark sequester'd pool,
> Among the fibres of the tree,
> The curious eye may often see
> A little crew of silver dace
> Self-prisoned in that shadowy place."—*Faber.*

The Dory, "with his dolphin head, whose fins like amber horns are spread" (Joanna Baillie), and whose body bears, so some say, the marks of Peter's fingers. (The disciple had caught it, but the fish cried so plaintively that, "touched with compassion," he picked it up and let it go, saying, "Rejoin thy family and remember me." Others, however, say in that notable passage of the river, when he carried the child-Christ across, St. Christopher caught the John Dory (a sea-water fish, by the way) and left the marks of the pinch which he gave it to be handed down *in memoriam* to the Dory's posterity. This fish had a certain classical sanctity as being called Zeus, and Aristotle has a "sacred fish," the Anthias, which, from his description of its habits, has been conjectured to be the John Dory. It was also called Faber, "the blacksmith," and so under the protection of Hephaistos, Mulciber, or Vulcan. Again, the Apah, or king-fish, is a native of the eastern seas, and it is not a little singular that, by a people so distant and secluded as the Japanese, this fish (originally included in the genus Zeus) should also be regarded as devoted to the Deity, and the only one that is so. The Apah is by them termed Tai, and is esteemed as the peculiar emblem of happiness, because it is sacred to Jebis or Neptune.)

"The insinuating eel" (Somerville), "lithe and wavy, in shining volumes rolled" (Pope); the "silver-throated" eels of Keats, "that dread the heron," and "weel-kenned for souple tail and gleds for greed," is, as it deserves, a prominent fish in verse, and many rivers strive for supremacy in the produce. Thus Pope has "the Kennet swift, for silver eels renowned;" in Drayton, "nor Annan's silvered eel exceedeth that of Trent ; "[1] and this creature may, therefore, lay fair claim to the honour of being a "time-honoured" dish, and its excellence a maxim of old prescription. (What fanatics the Greeks and Romans were about "the holy eel, Bœotian goddesses all clothed in beet!" They called it "the white-armed goddess" and "the Helen of the dinner-table." But the Egyptians went even further, for they not only adored them as food but formally worshipped them. None the less the natural history of the eel puzzled the ancients[2] vastly, and the theory that satisfied them best was that the creature was the spontaneous outcome of river-mud.)

The "disrelished" flounder "smooth and flat;" the spotted grayling "basking between the shadows" (Jean Ingelow), "whose great spawn is big as any pease" (Drayton).

"To smell daintily as a flower or a fish" has been accepted by our forefathers as an allowable simile. One angler says the smelt has a fragrance of lavender; another that it savours of cucumber; another that the grayling has the aroma of thyme. St. Ambrose called it the "sweet-flower of fishes." (The cuttlefish was supposed "with its sweet odour" to attract fish to it ; and the whale too obtained its food by opening its mouth, whence issued "so agreeable a scent" that the creatures of the deep gathered together in its jaws to enjoy the fragrant atmosphere. As a general rule too, the smell of fish cooking is considered rather worse than that of fish raw : yet, says an Athenian enthusiast, "the

---

[1] " Your Ulster eels," says Laud to Strafford (in Strafford's letters), " are the fattest and fairest I ever saw."

[2] Compare Pope's—

> " Silver eel in shining volumes rolled,
>   And yellow carp in scales bedropped with gold ; "

with Jenyns'

> " Here the bright silver eel enrolled
>   In shining volume lies,
>   There basks the carp bedropt with gold."

odour of a cooking conger is so divine that it would make a dead man sniff."

The mythical fragrance of the cuttlefish suggested to Domenichi to give the Cardinal Ferrara, as device, a sepia, with the motto, "Sic tua non virtus," meaning, that as the cuttlefish by its sweet odour attracts other fish around it, so the Cardinal, by the sweetness and affability of his disposition, drew all men after him—but not to eat them.)

The "daintie," "greedy," "fool-gudgeon" (in Crabbe, Moore, Cowper, and a dozen others a simile for dupes and victims of lawyers); the halibut, dignified by Cowper's verse as "the king o' the sea;" the herring, "sheathed in silvery green," moving "in crowds amazing," by many besides Swift and Darwin—

> " The migrant herring steers her myriad bands
>   From seas of ice, to visit warmer strands ;
>   Unfathomed depths and climes unknown explores,
>   And covers with her spawn unmeasured shores."

(Not that the herring attains in verse any prominence, adequate either to its intrinsic worth or its legendary importance. In folk-lore it is very conspicuous indeed, and in contemporary superstition no other fish, naturally enough, fills so large a space in the fancy of our sea-going folk, as may well be, if only from the Scotch fisherman's maxim, "No herring, no wedding." The connection may not at first be obvious to all, but the Scottish registers make it clear enough. In the returns for the third quarter of 1871, for instance, the registrar of Fraserburgh states that the herring fishery was very successful, and the value of the catch, including casks and curing, may be set down at £130,000 sterling, and the marriages were 80 per cent. above the average. On the other hand, the registrar of Tarbert has to report a steady falling off in the fishing at that creek, and consequently the quarter passed without an entry in the parish register. The registrar of Lochgilphead also returns that the herring fishery has been a failure in the loch, and states that this accounts for the blank in the marriage column that quarter—"No herring, no wedding."

That herrings have a king who leads them from sea to sea is a wide-spread tradition, but, as a rule, their monarch is some predaceous wretch who looks upon his subjects merely as a perambulatory larder;—thus the Chimæra, which in the north is called the herring king, does no more for the herrings than live upon them. The porpoise, by the way, is called both the "herring king" and the "herring pig;" but this creature's synonyms are singularly numerous.)

The "regicide" lampreys of Moore, that in Darwin, "with lungs respiring steer, kiss the rude rocks, and suck till they adhere," has only an indifferent character in folk-lore. So Gay—"Lamprey's a most immodest diet, You'll neither wake nor sleep in quiet."

The mackerel with its "tabbied sides," so delightful in Jean Ingelow's vignette—

> "And down we ran and lay upon the reef,
> And saw the swimming infants, emerald green,
> In separate shoals; the scarcely turning ebb
> Bringing them in; while sleek and not intent
> On chase, but taking that which came to hand,
> The full-fed mackerel and the gurnet swam
> Between."

"The minnows or spotted par with twinkling fin, swimming in mazy rings the pool within" (Joanna Baillie)—

> "The sudden rushing of the minnow shoal,
> Scared from the shallows by my passing tread."

"Dull as a mullet" was a Roman proverb, but in English verse it lives as a comparison for constancy, and is the "faithful" mullet—

> " But for chaste love the mullet hath no peer ;
> For, if the fisher hath surprised her pheer,
> As mad with woe, to shore she followeth,
> Prest to comfort him both in life and death."—*Walton.*

(In contrast to this, from the same poet, is the "sargus "—

> " The adult'rous sargus doth not only change
> Wives every day in the deep streams, but (strange)
> As if the honey of sea-love delight
> Could not suffice his ranging appetite,
> Goes courting she-goats on the grassy shore,
> Horning their husbands that had horns before."

So, too, the "constant" cantharis, which, says this "compleat" old angler,

> " In nuptial duties spendeth his chaste life,
> Never loves any but his own dear wife.")

The "bright-eyed perch with fins of Tyrian dye" (Pope), "and prickling fins against the pike prepared "—

> " As Nature had therein bestowed this stronger guard
> His daintiness to keep (each curious palate's proof),
> From his vile ravenous foe."—*Drayton.*

The pike that "feed like cannibals" (Butler); the "tyrant pike" (Somerville), "that keeps awful court" (Ramsay); " vile, ravenous pike," "dispeoplers of the lake" (Gay); the ruffe, "the very near ally" of the perch; the " wary " roach (Marvel), "whose common kind to every flood doth fall;" the "stately sawmont" of Burns, "that dreads the thievish net and triple spear" (Gay).

Only less frequent because less familiar is "the stately sawmont," for this noble fish meets everywhere with the same unstinted admiration. Its life-history might easily be gathered from the poets, but the two following must suffice as specimens of the detail in which "the lord of rivers" is treated.

" By purifying frosts when streams run clear,
  The amorous salmons to the fords repair ;
  Unerring instinct moves their longing mind,
  By wondrous ways to propagate their kind.
  Not the red firebrand blazing o'er their head
  Can force the lovers from their wat'ry bed ;
  So fierce love rages in their gelid blood,
  The unheeded trident gores them in the flood.
  Deep, deep, they bury in a sandy bed
  Their countless ova and prolific seed ;
  Which unobserved long lurk beneath the tide
  Till Sol arrays the year in vernal pride.
  Then all the sand (a true though wondrous thing)
  Begins to move as in a bubbling spring ;
  Swarming with life, the weltering bottom heaves,
  And glittering swarms crowd the encumbered waves.
  Broad shoals, on shoals in youthful prime, are rolled ;
  Their azure armour shines with studs of gold ;
  Bedropt with purple hues and scarlet bright,
  They shoot amid the floods, a glorious sight."—*Leyden.*

" Now when the labouring fish doth at the foot arrive,
  And finds that by his strength but vainlie he doth strive,
  His taile takes in his teeth ; and bending like a bow,
  That's to the compasse drawne, aloft himself doth throwe,
  Then springing at his height, as doth a little wand,
  That bended end to end, and flerted from the hand,
  Farre off itselfe doth cast ; so doth the salmon vaut.
  And if at first he faile, his second summersaut
  Hee instantlie assaies ; and from his nimble ring
  Still yarking, never leaves, until himselfe he fling
  Above the Weare."

The salmon-fisher too is a frequent figure, and Marvell's
vignette is perhaps the most curious—

" And now the salmon-fishers moist,
  Their leathern boatsbegin to hoist,
  And, like antipodes in shoes,
  Have shod their heads in their canoes.
  How tortoise-like, but none so slow
  These rational amphibii go."

The " dainty " sole ; the sturgeon of Darwin's poem—

" With worm-like beard his toothless lips array,
  And teach the unwieldy sturgeon to betray.
  Ambush'd in weeds, or sepulchred in sands,
  In dread repose he waits the scaly bands,

I

> Waves in red spires the living lures, and draws
> The unwary plunderers to his circling jaws,
> Eyes with grim joy the twinkling shoals beset,
> And clasps the quick inextricable net."

The "vengeful tench;" the trout "bedropt with crimson stains" in Shenstone, and "bedropped with crimson hail" in Burns; the weaver, "whose prickles," according to Drayton, " venom be;" whiting "in its pearly sheen" (Joanna Baillie), "a generall wholesome fish;" and all the rest—even the pout, loach, bleak, and bull-head, "which everywhere are found in every little beck "—find recognition in proportion to their deserts. The herring, salmon, and the trout enjoy, as is befitting, most frequent notice, and those who search such poets as Clare, Faber, Hurdis, Leyden, and Grahame, will chance upon passages of rare merit and prettiness.

The last-named fish is specially a favourite. It is "the river's pride."

> "The trout, by Nature marked with many a crimson spot,
> As though she curious were in him above the rest,
> And of fresh-water fish, did note him for the best."

Many poets were anglers, and so "the pools where the trout are found," "the rush-bordered rills where trout abide," "the birch-shaded place that harbours the trout," "the wimpling burn where darts the trout," are as familiar to readers of verse as the fish itself; "the wary trout that thrives against the stream" (Quarles); the "quick-eyed," "lightning-seizing" trout. Burns is alone, if I am not mistaken, in the idea of "death's fish-creel," making Death an angler. It would have been a finer fancy had the poet made him casting a *net* instead of a line; inasmuch as Death offers no lure to his victims.

CHAPTER III.

SOME POETIC FISH-FANCIES.

AMONG the foreign fishes—if any fish can be called
"foreign" to our islands—the shark, whale, sword-fish,
flying-fish, remora, nautilus, and "crimson" murex ("the
far-famed fish that gives the fleecy robe to Tyrian dye," and
"wounded strikes a purple stain"), afford the poets some
excellent touches of sea-life Nature.

The fiction of the "keeled" nautilus is far too beautiful
for any one to regret its survival, and the passages in which
our poets pass on to our time the legend of this mariner of
fancy are very dainty.   Montgomery's lines that begin—

> "Light as a flake of foam upon the wind,
>   Keel upward from the deep emerged a shell
>   Shaped like the moon ere half her horn is filled "—

are a delightful rendering of the myth, while Charlotte
Smith's poem to the nautilus is so charming that it is no
excessive generosity to forgive her adding a "tapering mast"
and "silken" sail to the rigging of the craft in which the
little mariner is supposed to "scud before the freshening
breeze "—

> "Where southern suns and winds prevail,
>   And undulate the summer seas,
>   The nautilus expands his sail
>   And scuds before the fresh'ning breeze.

> Oft in a little squadron seen
> Of mimic ships all rigg'd complete,
> Fancy might think the Fairy Queen
> Was sailing with her elfin fleet.
>
> With how much beauty is design'd
> Each channelled bark of purest white ;
> With orient pearl each cabin lin'd,
> Varying with ev'ry change of light.
>
> While with his little slender oars,
> His silken sail, and tapering mast,
> The dauntless mariner explores
> The dangers of the watery waste.
>
> Prepar'd, should tempests rend the sky,
> From harm his fragile bark to keep,
> He furls his sail, his oars lays by,
> And seeks his safety in the deep.
>
> Then safe on ocean's shelly bed
> He hears the storm above him roar,
> 'Mid groves of coral glowing red,
> Or rocks o'erhung with madrepore."

In another reference to the nautilus Montgomery repeats the same fancy—

> "Should a breath of danger sound,
> With sails quick furled it dives profound,
> And far beneath the tempest's path
> In coral grots defies the foe
> That never breaks, in heaviest wrath,
> The Sabbath of the deep below."

These are, of course, simply perpetuations of Pliny's engaging romance of the nautilus—one of the favourite traditions of the classical poets.

In this class of merely fanciful creatures may be also noticed the Pompilus, the sailor's pilot-fish, which was supposed to guide mariners to their destination, and having seen them safely into harbour, to go back to look for another job, for Apollo, it is said, changed a fisherman (named Pompilus) who had crossed him in his loves, into this fish, and condemned him for all eternity to the task of gratuitous

pilotage. The whale, again, was said to be attended by the
" musculus," a little fish that swam in front of Behemoth
and warned him off the shoals, on which he might have
otherwise run aground.[1] This legend reappears in the
" Pentameron," where the whale that has lost its way is told
to go and get " the sea-mouse " to pilot it.

The " remora," or sucking-fish, only finds recognition on
the score of the myth, that it could (as in Ben Jonson's
Magnetic Lady) stop vessels under full sail by suction.
Thus Spenser—

> " Looking far forth into the ocean wide,
>   A goodly ship, with banners bravely dight,
>   And flag in her topgallant I espied
>   Through the main sea making her merry flight.
>   Fair blew the wind into her bosom right,
>   And th' heavens looked lovely all the while
>   That she did seem to dance as in delight,
>   And at her own felicity did smile.
>   All suddenly there clove unto her keel
>   A little fish that men call remora,
>   Which stopped her course and held her by the heel
>   That wind nor tide could move her thence away.
>   Strange thing meseemeth that so small a thing
>   Should able be so great an one to wring ! "

In modern times it has been used to illustrate the power
of technical trivialities to retard a lawsuit, but antiquity believed it had the power of arresting a ship under full sail
by attaching its tail-end to a rock, and its head-end to the
keel of a passing vessel—

> " The lazy remora's inhaling lips
>   Hung on the keel, retard the struggling ships."

---

[1] Mallet on " The Critic " is excellently severe :—

> " Where'er the king of fish moves on before,
>   This humble friend attends from shore to shore
>   With eyes still earnest, and with bill inclin'd,
>   He picks up what his patron drops behind.
>   With those choice cates his palate to regale,
>   And is the careful Tibbald of a whale."

In the natural history of the period we read that "there is a little fish, keeping ordinarie about rockes, named Echeneis. It is thought that if it settle and sticke to the keel of a ship under water, the ship goeth the slower by that means, wherefore it is called the 'stay-ship.'" Now Pliny is here cautious enough, and attributes no more to the remora than is actually the property of barnacles when in number. But popular fancy outran fact, and a single remora, four inches long, was supposed to have held back Antony's flag-ship in the sea-fight off Actium. Periander also, among others, declared himself the victim of a similar accident, and the fiction flourished, thanks chiefly to poets and heralds, till a couple of centuries ago—

> " The sucking-fish, with secret chains
> Clung to the keel, the swiftest ship detains."

The flying-fish, "the aspiring fish that fain would be a bird," are, in the same way, favourites on account of a fancied power of protracted flight. Thus Leyden, in some lines written after being at sea for the first time—

> " On wondrous fins the fishes fly,
> Like birds on the ocean plain,
> In flocks like sparrows soar on high,
> And sport and glitter on the main."

while Shelley has the mysterious lines—

> " As the flying fish drop from the Indian deep,
> And mix with the sea-birds half asleep."

Montgomery, who often saw them, speaks of them flying "in bow-and-arrow" figures, which is accurate; but is he speaking from observation when he describes pelicans preying upon them when on the wing, "snapping them, as mosquitoes are by swallows"? Cowley correctly speaks of their "short silver wings," meaning of course that their flight is brief, and the metaphors taken from this brevity by Swift, Congreve, Shenstone and others, are both apt and forcible :—

> "So fishes, rising from the main,
>   Can soar with moisten'd wings on high ;
> The moisture dried, they sink again,
>   And dip their wings again to fly."

> "Sprung the flying-fish against the sun,
>   Till its dried wing relapsed from its brief height
> To gather moisture for another flight."—*Byron.*

> " Like those fish by sailors met,
> Fly highest while their wings are wet."—*Greene.*

> "Alas ! the flattering pride is o'er,
>   Like thee, awhile, the soul may soar ;
> But erring man must blush to think,
> Like thee, again, the soul must sink."—*Moore.*

The torpedo, "the benumbing fish," curiously meets with more references than might have been expected. An explanation, perhaps, is to be found in its familiarity to the ancients. A galvanic shock from the torpedo was one of Galen's prescriptions for rheumatics and the gout, and the poets, adopting the idea of antiquity, speak of the creature "delivering its opium at a distance"—the notion that electric fishes could discharge the fluid to a distance without any conducting medium. The spine of the stinging-ray was the barb of the spear that Circe gave her son, and to this day in the South Seas the savage tips his arrows and harpoons with them.

Sword-fishes, as the "warriors of the sea," lend a martial variety to many fishy lines, and we find them in Darwin in the company of "torpedoes, sharks, rays, and porpuses," pouring "their twinkling squadrons round the glittering shore."

> "Our little fleet was now engaged so far,
>   That like the sword-fish in the whale they fought ;
>   The combat only seemed a civil war,
>   Till thro' their bowels we our passage wrought."—*Dryden.*

The shark is in Nature a thing of exceeding horror, but the poets by their grisly fancies aggravate its terrors. Keats

calls them "thorny" sharks, and another poet "mailed;"
and it is a favourite thought—in Shelley, Montgomery,
Thomson, and Byron among others—that they fattened on
the slave-trade—

> " His jaws horrific arm'd with threefold fate,
>   Here dwells the direful shark.   Lured by the scent
>   Of steaming crowds, of rank disease and death,
>   Behold ! he, rushing, cuts the briny flood,
>   Swift as the gale can bear the ship along,
>   And from the partners of that cruel trade,
>   Which spoils unhappy Guinea of her sons,
>   Demands his share of prey—demands themselves.
>   The stormy fates descend ; one death involves
>   Tyrants and slaves ; when straight, their mangled limbs
>   Crushing at once, he dyes the purple seas
>   With gore, and riots in the vengeful meal."—*Thomson.*

> " Myriads of slaves, that perished on the way,
>   From age to age the *shark's* appointed prey,
>   By livid plagues, by lingering tortures slain,
>   Or headlong plunged alive into the main."—*Montgomery.*

How it follows ships when any are dying on board, is a
world-wide superstition, and finds place more than once in
verse.   Thus in Barry Cornwall—

> " Till dawn we watched the body in its dead and ghastly sleep,
>   And next evening, at sunset, it was slung into the deep !
>   And never, from that moment—save one shudder through the sea—
>   Saw we or heard the shark that had followed in our lee."

>   " Where
>   Is the ship ? On the verge of the wave where it lay,
>   One tiger is mingled in ghastly affray
>   With a sea-snake. The foam and the smoke of the battle
>   Stain the clear air with sunbows ; the jar, and the rattle
>   Of the snake's adamantine voluminousness,
>   And the hum of the hot blood that spouts and rains
>   Where the grip of the tiger has wounded the veins,
>   Swollen with rage, strength, and effort ; the whirl and the splash
>   As of some hideous engine whose brazen teeth smash
>   The thin winds and soft waves into thunder ; the screams
>   And hissings crawl fast o'er the smooth ocean streams,
>   Each sound like a centipede. Near this commotion,
>   A blue shark is hanging within the blue ocean,
>   The fin-winged tomb of the victor."—*Shelley.*

"With their hammocks for coffins the seamen, aghast,
  Like dead men the dead limbs of their comrades cast
  Down the deep, which closed on them above and around,
  And the sharks and the dog-fish their grave-clothes unbound,
  And were glutted like Jews with this manna rained down
  From God on their wilderness."—*Shelley.*

They are "the grim monarchs of insatiate death" (Darwin); and "tyrants of the flood," with "quenchless rage;" the "ravin'd salt-sea sharks." That they were supposed to have the snake's power—itself a fancy—of fascinating with the eye, especially women, and that on some coasts they are believed never to molest children, are superstitions which the poets utilise; and the use of their teeth for arming the terrible Polynesian cestus, is referred to by, I think, Thomson. The "shark and dog-fish" is a companionship that occurs with some frequency, notably in Shelley, and is not an unnatural one—

"As a shark and dog-fish wait
  Under an Atlantic isle,
For the negro ship whose freight
Is the theme of their debate,
  Wrinkling their red gills the while."

That the whale should be a familiar image in poetry is not remarkable, seeing how antiquity, Holy Writ [1] (as translated), art, heraldry, and fiction generally, have popularised Behemoth.

"Lo, warm and buoyant in his oily mail,
  Gambols on seas of ice the unwieldy whale;
  Wide waving fins round floating islands urge
  His bulk gigantic through the troubled surge;
  With hideous yawn the flying shoals he seeks,
  Or clasps with fringe of horn his massy cheeks;
  Lifts o'er the tossing wave his nostrils bare,
  And spouts pellucid columns into air;
  The silvery arches catch the setting beams,
  And transient rainbows tremble o'er the streams."—*Darwin.*

It is not often, however, that we hear the whale styled

---

[1] The word whale does not occur in the Old Testament.—P. R.

"grampus," as in Darwin's poem on the "Economy of Vegetation," in the striking passage that commences—

> "So when enormous Grampus issuing forth
>   From the pale regions of the icy north,
>   Waves his broad tail and opes his ribbèd mouth,
>   And seeks on winnowing fins the breezy south."

Nor is it easier to explain why poets should have persisted in thinking that whales lived on herrings—

> "A whale of moderate size will draw
>   A shoal of herrings down his maw;"

> "With monstrous gape,
>   Sepulchral whales devour
>   Shoals at a gulp;"—

than that they should say—three poets, at any rate—"white as whalebone." That it should be spoken of as "bounding" is not consonant with whale dignity; and when Mary Howitt adds to Montgomery's irreverence by saying bounding "*like lambs,*" it seems to the ordinary mind that a liberty is being taken with Behemoth, for even when at play the giant should be taken seriously.

> "With gills pulmonic breathes the enormous whale,
>   And spouts aquatic columns to the gale;
>   Sports on the shining wave at noontide hour,
>   And shifting rainbows crest the rising shower."

This treatment is seemly; and "cumbrous," "belching," "gulphing," "living islands," are all becoming epithets; but it is curiously significant of the common humanity that poets share with us that they should always exult so prodigiously in man's triumphs over the whale. Just as with the lion, nearly every time a poet talks of the whale "in oily mail," he is sure to say something about its being stranded, or harpooned, or hunted, or otherwise victimised, by human beings.

What an admirable poem that is of Mackay's on "The

Shoal of Whales." The fisher folk are all in chapel on a Sabbath when out through the open door one sees the monsters spouting in the bay. "There are whales in the bay! a shoal! a shoal!"

> "'Leviathan! Leviathan!'
> The minister cries, and shuts his book;
> And though a man of peace is he,
> As a preacher of the Word should be,
> He takes his musket from a nook,
> Rusty and old, and hastes away
> To join his people in the bay.
>
> The women have left their spinning wheels,
> Their hose, their nets, their fishing creels,
> And arm'd themselves with pikes and staves
> To follow the monsters of the waves.
> Fifty boats at least are ready,
> With rowers strong and helmsmen steady,
> To drive the whales into shallow water,
> And dye the beach with the blood of slaughter.
>
> Merrily row the boats,
> Merrily swell the sails,
> And merrily go the islanders
> To chase the mighty whales.
> And quietly prays the preacher
> For a blessing and reward
> Upon harpoon and musket,
> Upon the spear and sword,
> That shall slay the great Leviathan
> For the glory of the Lord."

But the Behemoth of the poets is, as I have said, a mixed beast. At times it is the elephant—"for thee his ivory load Behemoth bears." At others the rhinoceros—"decked in Behemoth's spoils, the tall Shangalla strode." And again a hippopotamus—

> "Mild is my Behemoth, though large his frame,
> Smooth is his temper and repressed his flame
> While unprovoked. This native of the flood
> Lifts his broad foot and puts ashore for food:
> Earth sinks beneath him as he moves along
> To seek the herbs."

But as a rule it is the whale.   So too Leviathan.   Though sometimes a crocodile, it is generally "that sea beast which God of all His works created hugest that swim the ocean streams," and which—

> " Hugest of living creatures on the deep,
>   Stretched like a promontory, sleeps or swims,
>   And seems a moving land, and at his gills
>   Draws in, and at his trunk spouts out the sea "—

Milton describes the whale as being often mistaken for an island :—

> " Him haply slumb'ring on the Norway foam,
>   The pilot of some small night-foundered skiff,
>   Deeming some island, oft as seamen tell,
>   With fixt anchor in his skaly rind,
>   Moors by his side, under the lee, while night
>   Invests the sea, and wished-for morn delays."

This island idea is similarly utilised by a score of poets. In another aspect Leviathan is a torpid giant of the true lazy giant breed :—

> "So close behind some promontory lie
>   The huge leviathans t'attend their pray,
>   And give no chase ; "

but tremendous and tyrannical when once in motion.   He is then "dread Leviathan," and "tempests o'er the main their terrors spread, to rock him in his bed."   From his "roaring nostrils" he sends "two fountains to the sky," and "spouts his waters in the face of day."   Sporting "on the face of the calm deep," "he lashes it into foam," and woe to those who "dispute his reign, and uncontrolled dominion of the main."

What manner of thing "Leviathan" was in those un-evoluted times—the period called (until the day of Lyell) the Epoch of Diluvium and Catastrophe, the age of unlimited mud—it would be almost profane for us, in these puny days of whales, without spirit enough in us to fish up even a sea-serpent, to attempt to imagine ; and for myself I am content to believe with the Talmudists that it was an undefinable sea-monster, of which the female lay coiled round the earth, till God, fearing her progeny might destroy the new globe, killed it, and that then He salted her flesh and put it away for the banquet which the pious shall enjoy at the Great End. In that day the angel Gabriel will kill the male also, and will make a tent out of his skin for the elect that are bidden to the banquet. But it is a hazy old tradition, I confess.

The " imperial whale " that does not dare, "unless by stealth," to attack the "firm united commonwealth of the herrings," is a very precious fiction, and full of humour.

> " But herrings, lively fish, like best to play
> In rowan ocean, or the open bay ;
> In crowds amazing through the waves they shine,
> Millions on millions from ilk equal line :
> Nor dares the imperial whale, unless by stealth,
> Attack their firm united commonwealth.
> But artfu' nets and fishers' wilie skill
> Can bring the scaly nations to their will."—*Allan Ramsay.*

Only a poet could imagine a whale stealing up in a red Indian, snaky sort of manner upon its prey, or suppose that Behemoth, let him be never so crafty, could take a herring by surprise, or fancy that any danger to the whales could possibly result from a bold front attack upon a shoal of herrings. That the whale eats herrings is "a fact of knowledge" with the poets, and seeing that they are convinced it really was "a whale" (as our translation of the Acts states), that "swallowed" Jonah, there can be obviously no physical difficulty in the way of Behemoth swallowing herrings. Pitt goes farther, and says it swallows sharks ! Milton and many others after him speak of the whale as being scaly, but Campbell is, I think, the only poet who endows Leviathan with a voice. "Where loud Lofoden whirls to death the roaring whale." Judging from that poet's natural history generally, it is more than probable that "roaring" was only a truth by accident, for it is undeniable that the whale and its cetaceous relatives, the "quadrupeds" of the sea, "can roar you" both "gently as any sucking dove," or "with that hollow voice of roaring" of the lion in the Progress that came after the Pilgrim at "a great padding pace." Our knowledge of these creatures is very insufficient, for, sad to relate, nearly all our reliable information of their habits has been obtained during the process of murdering them, or after the great things have been murdered. We have no acquaintance with the domestic or social life of the whales except in circumstances of terror and bloodshed. Under these conditions each of the sea-beasts have uttered sounds. The common dolphin makes "a murmuring, like suppressed lowing;" the other, while being killed, "bellowed like a bull." Porpoises when entrapped "uttered cries so distressing, that they had to be killed:" the grampus "groaned most horribly," and the "cries" of the "ca'ing whale" while being put to death were piteous. From all of which it may be assumed that in times of happiness the whale-folk are not voiceless; and there is abundant foundation for the belief of whalers, and the assent thereto of men of science, that the whales call to each other under water, and communicate the approach of danger by submarine signals of sound. So that the poets who include the whales among these fishes are incorrect in addressing them as "mute"—

> " Nay, the mute fish witness no less His praise,
> For those He made, and clothed with silver scales,
> From minnows to those living islands whales."—*Cowley.*

Nor indeed, when speaking of fishes proper, are poets always correct who address the finned things as "dumb" and "silent." For it is now established as a fact of knowledge that there are fishes who "pipe" and "snort" and "grunt," and who make noises that resemble drumming, whistling, and Jews'-harping.

The masted nautilus, again, is a very pleasant extension of the ancient myth, and suggests all kinds of possibilities—a double-funnelled nautilus, a passenger nautilus registered A1 at Lloyd's. Did any poet ever see one jury-rigged? The gleaming porpoise that "shoots" (in Moore) "along the wave," is a terror to mariners. Sea-horses "prick up their ears," and sea-dogs hunt prey "by scent." Seals, as we have seen, pursue children and pelt them with stones. Sharks kill their victims in Darwin "with descending blow," and in Shelley they have "red gills." That trout hybernate in river mud is as much a poetical fiction as Eliza Cook's whim that "pearls hang thick on the red coral stems;" or Congreve's vision of flying-fish in a river. Lobsters, according to one poet, live " in scarlet mail;" and the prawn when alive is, according to another, "pink-nosed."

The nautilus suggests to Montgomery a pleasant figure for the Esquimaux in their frail *kayáks*—

> " Trained with inimitable skill to float,
> Each balanced in his bubble of a boat."

The shark, of course, provides an easy synonym for all rogues of prey, and the gudgeon another for their victims:—

> " The roving eye, the bosom bare,
> The forward laugh, the wanton air,
> May catch the fop, for gudgeons strike
> At the bare hook and bait alike;
> While salmon play regardless by,
> Till art like nature forms the fly."—*Moore*.

Says Cowper—

> " Like trout pursued, the critic in despair,
> Darts to the mud, and finds his safety there."

"Live bullion" is Moore's pretty synonym for gold-fish;

which, by the way, are said to have first reached Europe as
a present to the Pompadour.

That eels grow in time familiar with the process of being
skinned, Byron and others jocularly note; and Pope's
"hold the eel of science by the tail" is only one of the
many allusions to the creature's slippery character and
evasive habits.

CHAPTER IV.

SOME "SHELL-FISHES."

CRUSTACEANS, "whose shell is all their bones," curiously enough bear a somewhat sinister character; but they have doubtless inherited this from antiquity, and so, having passed opprobriously into proverbs, became fair game for poets. At any rate, they do not hesitate to reflect upon the character of the lobster and the crab, and, as a sample, either Gay or Drayton is worth consulting by those interested in "heating" viands. Hood in his "Mermaid of Margate" celebrates the crustacean's revenge :—

> " The squealing lobsters that he had boiled,
>   And the little potted shrimps,
> All the horny prawns that he had spoiled,
>   Gnawed into his soul like imps."

Yet had our poets known as much as the Germans do of those monster crustaceans of Udvaer (which the fishermen dare not attack as their claws are a fathom apart) the poets might have been more respectful to the Jesuits' symbol. The crab, "as bargemen wont to fare, bending their force contrary to their face" (Spenser), is ridiculed for being "sidelong" and "crouching :"—

> " No crab more active in the dirty dance,
>   Downward to climb, and backward to advance" (*Pope*)—

and therefore papistical.

Next, "glowing" Cancer :—

> " As close in's shell he lies, affords his aid
> To greedy merchants, and inclines to trade."

But over births his influence is hardly more auspicious than the fishes', though in omen it is happy—

> " The dream's good ;
> The crab is in conjunction with the sun."

And it is by the Gate of Cancer, Mercury standing at the starry portals, that souls descend to take possession of the bodies of men. Not that the reasons of the crustacean's exaltation commend it to popularity, for when Hercules was fighting with the Hydra, Juno meanly sent Cancer to bite the hero's heel, but Hercules merely stopped for a moment in his job, killed the crab, and then went on with the Hydra. The goddess, however, translated the smashed crustacean to the skies.

In the story of the "Cruel Crane Outwitted," the bird, finding the fish likely to die of drought in a fast shrinking puddle, offers to carry them across to a large and pleasant lake of which he knows. After much suspicious demurring, the fishes go with the crane, one by one, and are, of course, eaten up in succession. Left last of all, however, is an old crab, and the bird proposes to take it over too, to join its old comrades. "Very good," says the crafty crustacean, "but as you cannot very well hold me in your beak as you did the fishes, suppose I hold you with my pincers." The crane agrees to this, and having arrived at the shambles, announces to the crab that he is now about to be eaten. "Not a bit of it," is the reply. "On the contrary, if you do not take me to the lake at once, I shall nip your head off your thin neck." So the crane in great alarm takes Cancer straight to the lake, but before getting off the bird's back, the crab bites its head off.

And, so again, when he kills the snake and sees it lying stretched out along the ground, he addresses the dead viper with the caustic moral—"This fate would never have befallen you if you had lived as straight as you have died."

The crab runs the fox a race, and as soon as his opponent starts, catches hold of its tail. When the fox reaches the winning-post it turns round to see how far the crab has got, when the wily crustacean quietly drops off, crosses the winning line, and startles the fox with—"What! come at last, are you? I've been here some time."

> "The scallop cordiall judged, the dainty whelk and limp,
> The periwinkle, prawnes, the cockle and the shrimp,"

are all of ill-repute of being what a poet calls "a most immodest diet." (In Gay's poem "To a Young Lady" these traditions are all noticed.) "The lobster, as an enemy to serpents, was," says Moule, "sometimes used as an emblem of temperance, and two lobsters fighting as an emblem of sedition." The union of a lobster with the human form is an impresa of very old date, but the families on the continent that bear this crustacean for a badge probably refer it back to no earlier times than the chivalric days when knights went forth to fight in that armour of overlapping plates which were called "ecrivisses." But just as many have adventitiously arrived at honour, so many others have accidentally fallen into disrepute, and the lobster, recollecting its traditional obliquities, may hardly go haughtily. Its character in legend is very curious, for, while the crab is always of good repute, the lobster is ever of bad. Very old engravings show us a fool astride a lobster, and the significance of that medal of the Pretender, in which the youthful aspirant is shown in the arms of a Jesuit who rides a lobster, conveys nothing to the credit of either the friar or the "fish." Mercury in his baser aspect rides a cray-fish. Prawns and shrimps are among the heraldic bearings of the Crafords and Atseas of Kent; and the cray-fish, also an English crest, was the badge of the Prince of Orange, and betrayed that warrior to imprisonment when he had hoped to escape identification among a heap of the killed after the battle of St. Aubin du Cormier. The crab frequently recurs in heraldry—the golden crabs of the Scropes, the Danbys, and the Bytheseas.

A scallop on a shield shows, or should show, that an ancestor had been in the Crusades, as it was the cognisance of St. James, and after him of all who fought against the infidels, and so of all pious pilgrims. The badge of the Order of St. James of Spain is a sword with a cross handle and a scallop on the pommel. The same shell forms the badge and collar of the Order of St. James in Holland, and St. Louis instituted the "Order of the Ship and Escallop" for the decoration of the nobility who accompanied him to the Holy Land. The collar of the Order of St. Michael, founded by Louis XI., was garnished with golden scallops. The cockle, whelk, and several of the genera Turbo and Cyprœa, found among modern crests and shields, date back to the palmy days of Phœnicia, when Tyre and other cities of the Mediterranean stamped their medals and coins with them.

But of course the most celebrated and popular shell crest and device was the pearl-oyster. Charged with its precious freight, it appears in a hundred forms, the legend always repeating one or other of the curious and beautiful fancies of antiquity. Every royal ·Margaret, by right of name, claimed the precious thing as her emblem, Princes and nobles bore it on their impresas, and the coronets of nobility take their degrees of rank from the pearls upon them.

The poets seem hardly to have done justice to these beautiful houses tenantless, the shells, which once held the happy creatures of the deeps. "Buoying shells, that on stormless voyages, wherried their tiny mariners." "Pink-lipped and rosy," sometimes "humming," and one poet has the line, "all the wonders of the cockle-shell." Yes, they are wondrous things these gifts of the sea, strewn upon our beaches as hints of hidden treasures :—

> "Gold, amber, ivory, pearls, ouches, rings,
> And all that else was precious and dear;
> The sea unto him voluntary brings."—*Spenser.*

Then, too, that beautiful cave-recalling whisper of the shells :—

> "Apply
> Its polished lips to your attentive ear,
> And it remembers its august abodes;
> And murmurs as the ocean murmurs there."

And again :—

> "I have seen
> A curious child who dwelt upon a tract
> Of inland ground, applying to his ear
> The convolutions of a smooth-lipped shell,
> To which, in silence hushed, his very soul
> Listened intensely; and his countenance soon
> Brightened with joy; for murmurings from within
> Were heard, sonorous cadences! whereby
> To his belief, the monitor expressed
> Mysterious union with its native sea."—*Wordsworth.*

These are fine passages. But for most of the poets the shell and its occupant denote only the "lowest" form of life and intelligence. "From the mute shell-fish to man," touches the extreme poles, and Pope in conchology finds the lowest deeps of dulness :—

> "Yet by some object every brain is stirr'd,
> The dell may waken to a humming-bird;
> The most recluse, discreetly open'd, find
> Congenial matter in the cockle-kind."

In contrast is Somerville's admiration, albeit tempered by a pitiful "poor" and a qualifying "even :"—

> " The curious sage
> Ranks in classes all the fishy race,
> From those enormous monsters of the main,
> Who in their world like other tyrants reign,
> To the poor cockle-tribe, that humble band,
> Who cleave to rocks, or loiter on the strand.
> Yet ev'n their shells the forming hand divine
> Has, with distinguish'd lustre, taught to shine.
> What bright enamel, and what various dyes,
> What lively tints delight our wond'ring eyes."

What reckless adventurer first dared to swallow an oyster? Was it some miserable pearl-diving slave who desperately hoped to steal the accompanying gem, and thus arrive at the price of his ransom?

> " For them the Ceylon diver held his breath,
> And went all naked to the hungry shark,
> For them his ears gushed blood."—*Keats.*

Or was it some poor shipwrecked wretch, groping about among the rocks of his barren island, who, to save dear life, first bolted the unlovely bivalve? Surely nothing less than a last despairing effort to regain liberty or preserve life could have nerved the primitive oyster-eater to the shocking act. The poet Gay attributes the discovery to the jaded appetite of luxury.

> " The man had sure a palate covered o'er
> With brass or steel that on the rocky shore
> First broke the oozy oyster's pearly coat,
> And risk'd the living morsel down his throat."

" What will not luxury taste ? " he asks. It does not follow from this that Gay did not like oysters; indeed there is evidence in his verse that he was partial to shell-fish. But the tendency of poets is to make fun of the amiable mollusc. Cowper, who, it seems to me, had very little real tenderness for the weak, though something of a jelly-fish himself, sneers at the oyster for not feeling anything till " the knife is at its throat." Byron laughs at the idea of its being crossed in love. And it is a fact that, in spite of classical traditions as to the excellence of the creature as food, our ancestors held the oyster in disrepute. "Not worth an oyster," was a common saying in Chaucer's time, and the Sompnere expects sympathy from his audience inasmuch as he was driven to such straits for food that he actually had to eat oysters. "For many a muscle and many an oistre, When other men have been ful wel at ese, Hath been our food." Nowadays but little sympathy, I imagine, would attach to a Londoner compelled by circumstances to an occasional oyster diet.

Drayton mantains the excellence of the native :—

> " Thinke you our oysters here unworthy of your praise,
> Pure Wall Fleet which doe still the daintiest pallats please,
> As excellent as those which are esteemed most."

We have all read how it was a tradition, and not so very long ago, that barnacle-geese were bred from the barnacle. Says Butler, "As barnacles turn Soland geese, in the islands of the Orcades."

> " Whereas those scattered trees, which naturally partake
> The fatness of the soyle, send from the stocky bough
> A soft and sappy gum, from which these geese do grow
> Called barnacles by us, which like a jelly fish
> To the beholder seeme, then by the fluxure moist
> Still great, and greater thrive untill you well may see
> Them turn'd to perfect fowles, when dropping from the tree
> Into the meery pond, which under them doth lye,
> Wax ripe, and taking wing away in flocke doe flye,
> Which well the ancients did among our wonders plaie."

CHAPTER V.

THE POETS' DOLPHIN.

A THOROUGHLY poetical fish is the poets' dolphin.   In the
first place, it is not a dolphin at all, but a porpoise.   In the
next, it is a creature abounding in classical impossibilities,
and as overgrown with delightful legends and traditions as
any old tree in an Assam jungle is with orchids.   Moreover,
it is a "fish" the poets really know nothing about.   Finally,
as both Arion and Amphion, the heroes of delphinian
adventure, were poets, their successors of more modern
days feel themselves called upon to patronise the dolphin.[1]
It is therefore in every way a thoroughly poetical animal.

Physically, it is "scaled" and "curved" in form, while
its colours transcend all other efforts of nature in loveliness.
It plays in "coral caves," and lives in "caverns roofed with
spar."   Sometimes it comes forth to do battle with the seal,
as in Spenser :—

> "As when a dolphin and a sele are met
> In the wide champain of the ocean plain,

---

[1] Davenant is a delightful exception.   Speaking of the Drum he says :—

> "Had wet Arion chosen to lament
> His grief at sea on such an instrument,
> Perhaps the martiall mustick might incite
> The sword-fish, thrasher, and the whale to fight,
> But not to dance."

> With cruell chaufe their courages they whet
> The maysterdome of each by force to gaine,
> And dreadfull battaile twixt them do arrayne ;
> They snuf, they snort, they bounce, they rage, they rore,
> That all the sea, disturbed with their traine,
> Doth frie with fome above the surges hore :
> Such was betwixt these two the troublesome uprore."

Morally it is a sportive fish of very benevolent tendencies. It warns sailors of the approaching storm, guards their keels,[1] and befriends the mariner at sea in every way it can. Poets indeed would have been grievously put to it for a touch of marine nature, if it had not been for this "fish" they call the dolphin. As a rule, the poets mean the porpoise, but it is a fair exercise of poetical license to say dolphin for porpoise, or in fact for fish generally. For this creature stood in ancient symbol and classic verse for "the sea-things," so modern poets always keep a dolphin on hand, in case their wanderings in search of a rhyme should take them to the sea coast. It is in fact a stock-in-trade "fish," like the nautilus and the shark and "Behemoth," just as the owl, dove, eagle, lark, and "night-raven" are stock-in-trade birds, or the tiger, fawn, wolf, fox, and sheep stock-in-trade beasts. Every legend of the mystical thing has been punctually utilised by our poets, and often with great elegance and force. Was not the dolphin "the king of the sea," the symbol of marine sovereignty, emblem of sea cities and their maritime supremacy, the hieroglyphic of Neptune himself? So the poets delighted in "these fleetest coursers of the finny race," the synonyms of celerity, the special messengers of sea-divinities.

And what delightful creations of fancy they are, those "Ionian shoals of dolphins that bob their noses through the brine" in Keats! Here in Darwin, one "playful draws

---

[1] Thus in Waller :—

> "About the keel delighted dolphins play,
> Too sure a sign of seas' ensuing rage."

Galatea's silver shell," and "she strikes the cymbal as it moves along."   Here another in Montgomery, "enamoured of the Triton's music," floats attentive.   A third carries the crowned Arion—

> "Who playing on his harpe, unto him drew
> The eares and hearts of all that goodly crew;
> That even yet the dolphin, which him bore
> Through the Ægean seas from pirates vew,
> Stood still by him, astonisht at his lore,
> And all the raging seas for ioy forget to rore."

Spenser says, "Through the sweet musick that his harp did make, allured a dolphin him from death to ease."   A fourth, as the willing steed of Apollo, "to delight his ear doth load his back" (Davenant).   On the one hand, with the older poet, we see "a team of dolphins, ranged in array, draw the smooth chariot of sad Cymoent;" on the other, with the modern Amphitrite in her "pearly car" (Wordsworth).   "Upon the crooked dolphin's back, scudding amidst the purple-coloured waves" (Green), and "Theban Amphion, leaning on his lute as he skims the sea" (Keats). Thus, too, the melody of "sweet shells" welcomes returning Thetis, but the goddess beams with greater pleasure upon "the dolphin tumults" that herald her approach. They are the playmates of Endymion, letting him "feel their scales of gold and green;" the bright herds of Proteus; "the dolphin faye" that "won Deucalion's daughter bright" (Spenser); the "gen'rous dolphins" that "in sportive ring" guard Neptune's statue (Savage).

How often these images of antiquity are reproduced by courtier bards I will not venture to say; but, as a rule, whenever kings and queens are on the seas, "about the keel delighted dolphins play;" and if one poet wishes to pay another a compliment, he speaks of "the gift to King Amphion, that walled a city with its melody."

So, too, Proteus has his modern representative in him

of Fingal's cave, where "by turns, his dolphins, all finny palmers great and small, came to pay devotion due."

"Crooked" is a favourite poet's epithet for the dolphin, for most poets go to heraldry and fable for their natural history ; and antiquity always spoke of the dolphin, drew it in pictures, carved it in stone, and engraved it on metal, as curly-bodied. Sometimes they were merely bowed "convexedly ;" in others, as in Arion's steed, "concavously inverted," with various spinal inflexions. But this incurvity is, of course, a fiction, for the dolphin is as straight-spined an animal as ever swam in water ; but the popular error rose, not unnaturally, from the fact of the porpoise being usually seen when leaping, and with its head declining towards the water before the tail had fairly left it. Among the pretty fancies which the evanescent beauties of the dying dolphin have afforded poets, should be quoted Byron's, of the parting day, that—

> "Dies like the dolphin, whom each pang imbues
> With a new colour as it gasps away,
> The last still loveliest till 'tis gone, and all is grey."

Anticipated, however, among others, by Falconer in the lines—

> "While in his heart the fatal javelin thrills,
> And flitting life escapes in sanguine rills ;
> What radiant changes strike th' astonished sight,
> What glorious hues of mingled shade and light !
> Not equal beauties yields the lucid west,
> With parting beams all o'er profusely drest,
> Not lovelier colours paint the vernal dawn,
> When orient dews impearl th' enamelled lawn "—

and Herbert's characteristic lines—

> "Oh ! what a sight were man if his attires
> Did alter with his mind,
> And like a dolphin's skin,
> His clothes combined with his desires."

Not that Herbert keeps his metaphor within an allowable

licence any more than Montgomery, who speaks of the dolphins when alive as miracles of beauty. They "glow with such orient tints, they might have been the rainbow's offspring," and "sparkle through the sea." Other poets make the same fanciful error, and thus it may be that Eliza Cook spoke of them as "golden."[1] It is a standard of comparison for the gaiety of others :—

> "The longing heart of Hero, much more joys
> Than nymphs and shepherds when the timbrel rings,
> Or crooked dolphins when the sailor sings."

> "And when he chose to sport or play,
> No dolphin ever was so gay."

"Sportive," "leaping," "gay," are among the regular epithets of the creature, and thus a poet arrives at the delightful phrase of "the dolphin Pleasure."

The porpoise then, up to a certain point, is the counterpart of the "pellochs rolling from the mountain bay"—the dolphin. Nor, seeing that the poets really, as a rule, meant porpoise when they said dolphin, is this to be wondered at. Both, therefore, are "gamesome," "tumbling," and so forth.

But here comes in one of those ingenuities of fancy so characteristic of poetical natural history—the creation of two species out of one. For the porpoise has many disadvantages as compared with the dolphin. Its name is absurd; it is never mentioned in classical verse; it had no taste for music; never happened to be near a ship when poets were thrown overboard with their harps; never dragged goddesses about in shells. This being the case, the poets deliberately transfer its virtues, under the name of dolphin, to an imaginary creature, and reconstruct a new fish of fancy out of what is left. This is exactly analogous to the other poetical process of crediting the

---

[1] So Keats, following the Greeks' epithet of "gleaming," speaks of their "scales of gold."

eagle and the pelican with all the virtues which antiquity really intended for the vulture, and then abusing the vulture as a "gryffe" or a fowl without any virtues.

Though gamesome like the dolphin (that is itself), the porpoise is also ferocious "like as a mastiff"—

> " Upon every side
> The gamesome porpoise tumbled in the tide,
> Like as a mastiff, when restrained a while,
> Is made more furious and more apt to spoil" (*Quarles*)—

while the extension of the weather-foretelling idea into something sinister, is a curious illustration of the poet's humour. That "the tumbling porpoise" predicted rough weather was an old sailor's superstition before Virgil's time, and is a familiar idea in English verse, as in Churchill, "like a porpoise just before a storm onward he rolled," or in "The Shipwreck"—

> " When threatening clouds th' etherial vault deface
> Their route to leeward still sagacious form,
> To shun the fury of th' approaching storm."

But it ought to have been noted by other poets that the porpoise foretold coming rough weather, by its anxiety to escape from it, and not from any delight in the tempest. For the only use of the creature as a weather-gauge was the direction in which it was going, inasmuch as the porpoise is always seen hurrying away in front of the storm which it apprehends to be coming. Churchill and Falconer understood this, but the majority simply made the poor porpoise "the vulture of the sea," delighting in conflicts, and eating the victims of the storm. Thus in Davenant :—

> " The prince could porpoise-like in tempests play,
> And in court-storms on shipwrecked greatness feed."

Strictly analogous also to the poetical evolutions of a specially sinister "night-raven" out of the ordinary raven, is the poets' deduction of a "porcpisce" out of the ordinary porpoise, which they treat as if it were a pig, and call it "greedy" and "obscene."   The grampus is a "variety" of the porcpisce which poets decline to entertain seriously—except when they mean a whale.

The following note of Wren's upon the use of the word "porpoise" by Sir Thomas Browne is worth quoting :—

"Read porkpisce.   The porkpisce hath his name from the hog hee resembles in convexity and curvitye of his backe, from the head to the tayle ; nor is he otherwise curlie, than as a hog is ; except that before a storme hee tumbles just as a hog runs.   That which I once saw cutt up in Fish Street was of this forme, and above five foote longue, his skin not skaly, but smoothe and black, like bacon in a chimney, and his bowels in all points like an hog ; yf instead of his four fins you imagine four feete, hee would represent a back hog (as it were) sweated alive."

"They say they are half-fish, half-flesh," says the prince in "Pericles," and one of the many legends of the porky nature of this fish to which the poets subscribe is that St. Patrick, unable to restrain his appetite on a fast-day, surreptitiously procured some pork and proceeded to eat it. But an angel, to save the saint from sin, turned it into porpoise meat, and so the porpoise has been partly pork ever since.

PART III.

THE POETS' INSECTS.

CHAPTER I.

ANTS AND BUTTERFLIES.

AMONG the morals and maxims drawn from natural history few are so popular, or so unfair, as that which contrasts the "industrious" ant with the "idle" butterfly. It requires only the most rudimentary knowledge of insect life to understand this. For the butterfly, after all, is only the winged caterpillar. In a previous stage it worked with monstrous diligence at the work it was given to do, consuming every day ever so many times its own weight of green stuff. Then it went into its chrysalis state and rested for a while and changed itself, and by-and-by it came out of its shell, put on its gay wedding clothes and flew away to find a mate. The honeymoon over, the female set about laying her eggs, and it was probably when it is engaged in this very serious business that the charge of idleness has attached to it.

For the insect flutters about the plant which it knows to be the proper food for its young, and settles for half an instant here and half an instant there, its whole demeanour being frivolous and flighty. Such, of course, is not the truth, for every time that the butterfly rested an egg was laid, and while it fluttered, apparently so aimlessly in all directions, it was really in its instinct selecting proper cradles.

At other times, too, when flitting so "idly" from flower to flower, it was just as busy as the bee which shared the flowers with it, for it was on the same business, of eating honey.   So that as a matter of fact even the winged butterfly has its work to do, and does it conscientiously and thoroughly.   The winged ant does no more; indeed, it does much less, for it has not even to feed itself, nor has it to take any maternal care of its young whatever.   All this is done for it.   Its only duty is to get married, and as soon as this is done, it is taken possession of by the unwinged workers, and spared all further exertion thenceforward from that moment.

Nor if we compare the unwinged butterfly, the caterpillar, with the wingless ant, is the balance of work in favour of either.   Each gets through all that it has to do with exemplary industry and despatch.

Amongst ourselves, too, there are the butterflies and the ants—the men who get through quantities of work and yet somehow always seem to have leisure for enjoyment, and the men who moil and toil all the day long, grinding out work and complaining that the day is too short for all that they have to do.   Yet the actual output of the former is usually as large, and as a rule more brilliant, than that of the latter.   Moreover, he goes about his work—and in the intervals of his work about his pleasures—like a creature that is capable of happiness, a pleasure to look at, understanding the gladness of sunshine and the beauty of flowers

The ants among us wonder at them, and in their clumsy antennal language exchange astonishment over their butterfly fellow-workers being able to get through so much and yet seem so unoccupied.   They themselves can hardly find time to take their meals, or make themselves tidy, and hurry about in the heat and dust dragging loads behind them, shabby and careworn, while the butterflies carefully dressed flit from work to play, all gaiety and good temper, and with time to spare for a laugh and a gossip.

Watch two ants meet on a stick, as I did the other day. One had in tow half a fly, with a pair of legs and one wing on it, and the trouble that these limbs gave the ant, catching in everything as it went along, was perfectly exasperating to look at.

But the ant tugged and lugged at the thing with stupid, dogged determination, and so, going round and round, upside down and downside up, it slowly, painfully moved on its silly way. The other had got a grass seed in its pincers, and the husk was certainly one half of the total weight : yet this " wise emmet " was going to haul this grain to its nest, and right down to the bottom of it, then husk it and laboriously drag the useless outside cover up to the front door again and throw it away.

Each of the two, in fact, was doing exactly twice as much as there was any necessity for doing : was wasting half its time and fifty per cent. of tissue in accomplishing a perfectly useless task of industry.

Well, they met on a narrow stick, and I could have banged their two heads together for the way they went on. In the first place, why were they on the stick at all? It was very narrow, very crooked, and not steady. The ground on either side was not only easy and open going, but a much shorter way. However, there they were, and they met.

The ant with the fly was going backwards, tug—tug— tugging at its inconvenient treasure. The legs of the half-fly hitched in every crevice of the wood, the wind kept on catching its one wing, and nearly capsizing the ant. It was just as if a man should try to drag, say, a dead camel through a jungle with a balloon tied to its head, and a couple of fifty pronged anchors tied to its tail. Travelling the opposite way was the other ant with its grass seed in front of its face, so that it could not see an inch before it. And so the two collided.

Now I think, from personal observation, that no other insect is more unphilosophically given to sudden panic than the ant. If anything out of the ordinary happens, the scene is one of instant and utter demoralisation.

By-and-by instinct reasserts itself, and they combine to retrieve the disaster which the commonwealth may or may not have suffered. But at first it is all frenzy. The most intimate of friends turn on each other in mortal combat, and when the mistake is discovered, rush off in opposite directions in search of fresh adventures without as much as a "beg your pardon" between them.

The present case was no exception. On the instant that the blindfolded ant ran into the tail of the other, both decided that there was nothing for it but immediate and precipitate flight. Each dropped the burden which it had been carrying with such furious pertinacity, and falling simultaneously off the stick whirled away as fast as its legs could carry it.

Even in the ordinary rencontres of everyday life they are aggravating and disagreeable. Hurrying round a corner two of them knock their heads together. " Hullo," says each of them, and then they stand opposite each other with nothing to say. All of a sudden both perk up and say simultaneously, "Well, good bye ! I've got to go to catch a train," and hugely relieved at getting away from each other, spin along the road as if the police were after them. Yet they are perfectly honest insects, these, only they have no idea how work ought to be done, or that there is anything in life beyond the sordid routine of getting through a job, and, if possible, saving up. All they know about sunshine is that it makes them sweat.

" Flowers ! Yes, pretty things flowers. My little girl is very fond of flowers," said an old stupid to me one day in the train.

But the butterflies—what gentlemen they are. What an

educated sense of beauty they have, and what inherited taste. Who, on the one hand, has ever seen a butterfly that did not do its work properly, or, on the other hand, was in such a hurry that it had not time to be civil. I know that butterflies sometimes develop sordid and morbid likings for raw meat and other abominations, but all the same I think that, taking the insect-world all round, there is no other class that fulfils its duties with, at once, more credit to itself and more pleasure to others than the "idle" butterflies. They seem only an ornament of society, and yet, if they were gone, how substantial would be their loss ! The ants, doubtless, do not think so, for the formic proletariat look upon the necessity for labour as a brand—an ignoble birth-mark, and yet imagine that the measure of an insect's worth is in the sweat of its limbs.

And now I have gone beyond the poets in imagining; but let that be as it may—I have not gone beyond the facts of nature. Ah me! the fables that the hedgerows tell; the parables of the fields ! Of these the poets have known little. Imagine even Thomson writing this fustian :—

> " Behold ! ye pilgrims of this earth, behold !
> See all but man with unearn'd pleasure gay :
> See her bright robes the butterfly unfold,
> Broke from her wintry tomb in prime of May !
> What youthful bride can equal her array ?
> Who can with her for easy pleasure vie ?
> From mead to mead with gentle wing to stray,
> From flower to flower on balmy gales to fly,
> Is all she has to do beneath the radiant sky."

And thus Error has come down to us in their lines, sanctified by the beauty of their genius and the magic of their speech.

> " As an ant of his talents superiourly vain,
> Was trotting with consequence over the plain,
> A worm, in his progress remarkably slow,
> Cry'd—" Bless your good worship, wherever you go !
> I hope your great mightiness wont take it ill,
> I pay my respects with a hearty good will !"

With a look of contempt and impertinent pride,
"Begone, you vile reptile," his antship reply'd;
"Go, go, and lament your contemptible state,
But first, look at me, see my limbs how complete!
I guide all my motions with freedom and ease,
Run backward and forward, and turn when I please.
Of nature (grown weary) you shocking essay!
I spurn you thus from me, crawl out of my way."
The reptile insulted, and vexed to the soul,
Crept onwards, and hid himself close in his hole;
But nature, determin'd to end his distress,
Soon sent him abroad in a butterfly's dress.
Ere long the proud ant, as repassing the road,
(Fatigu'd from the harvest, and tugging his load)
The beau on a violet bank he beheld,
Whose vesture, in glory, a monarch's excell'd;
His plumage expanded —'twas rare to behold,
So lovely a mixture of purple and gold.
The ant, quite amazed at a figure so gay,
Bow'd low with respect, and was trudging away:
"Stop friend," says the butterfly, "don't be surprised;
I once was the reptile you spurn'd and despised;
But now I can mount; in the sunbeams I play;
While you must for ever drudge on in your way!"

Moral,

" A wretch, tho' to-day he's o'erloaded with sorrow,
May soar above those that oppress him to-morrow."

Allan Ramsay, with a poet's usual fidelity to originals, has
a poem with the same title and identical ideas as this of
Cunningham's. It commences—

" A pensy ant right trig, and clean,
Came one day whidding o'er the green;
When to advance her pride she saw
A caterpillar moving slaw."

The caterpillar is very civil to the ant who, however,
"gecks up" her head and addresses it in scorn as "Some
experiment of nature, Who scarce has claims to be a
creature," and boasts of her own superior person:—

> " For me, I'm made with better grace,
>   With active limbs and lively face,
>   And cleverly can move at ease,
>   Frae place to place where'er I please."

The caterpillar goes off humbly into retirement, and the end is just as in Cunningham's poem.

In Parnell, for instance, the flies as "foolish nurslings of the summer air" are violently contrasted with ants as wise creatures :—

> " Black ants in teams come dark'ning all the road,
>   Some call to march and some to lift the load :
>   They strain, they labour with incessant pains,
>   Press'd by the cumbrous weight of single grains.
>   The flies, struck silent, gaze with wonder down ;
>   The busy burghers reach their earthly town,
>   There lay the burthens of a wintry store,
>   And thence unwearied part in search of more."

Somerville also has a singularly infelicitous poem on the same subject, when we remember that the poor ants die about the same time as butterflies :—

> " The careful ant that meanly fares,
>   And labours hardly to supply,
>   With wholesome cakes and homely tares
>   His num'rous working family.
>
> Upon a visit met one day
>   His cousin fly in all his pride,
> A courtier insolent and gay,
>   By Goody Maggot near ally'd.
>
> The ant who could no longer bear
>   His cousin's insolence and pride,
> Toss'd up his head, and with an air
>   Of conscious worth he thus reply'd.
>
> ' Vain insect ! know the time will come
>   When the court sun no more shall shine,
> When frosts thy gaudy limbs benumb,
>   And damps about thy wings shall twine ;
>
> When some dark nasty hole shall hide,
>   And cover thy neglected head ;
> When all this lofty swelling pride
>   Shall burst, and shrink into a shade.

Take heed lest fortune change the scene :
  Some of thy brethren, I remember
In June have mighty princes been,
  But begged their bread before December.' "

But Charlotte Smith, as always, swings the matter back into proper grooves :—

" Beneath some leaf of ample shade
Thy pearly eggs shall then be laid,
Small rudiments of many a fly :
While thou, thy frail existence past,
Shall shudder in the chilly blast
And fold thy wings and die.

Soon fleets thy transient life away,
Yet short as is thy natal day
Like flowers that form thy fragrant food ;
Thou, poor Ephemeron shalt have filled
The little space thy Maker willed,
And all thou know'st of life be good."

But to descend from great to small, from the general to the specific. I am afraid, much afraid, that poets thought *that ants made mole-hills.* I tremble while I write it, but did Wordsworth himself (one of the most inaccurate and unsympathetic of observers) think so ?

" So the emmet gives
Her foresight and intelligence that makes
The tiny creatures strong by social league,
Supports the generations, multiplies
Their tribes, till we behold a spacious plain
Or grassy bottom, all, with little hills
Their labour covered, as a lake with waves
Thousands of cities, in the desert place
Built up of life, and food and means of lif·."

Or, at any rate, that ants were naturally associated with them. Curious theories of zoological relativity are not uncommon with poets, as, for instance, among birds, the connection between doves and vultures, or among beasts, between the wolf and the bear, or among reptiles, between newts and adders. None the less, the ant-and-molehill association is sufficiently original. Did the poets really think that ants made mole-hills? It looks like it.

Thus, we read in " Quarles," that man is "a pismire crawling on this mole-hill earth," and if we analyse the cerebration here, we can only conclude that the poet had in his mind some vague idea that the industry of the pismire had erected the " hill "—which he calls a " mole-hill." King says, "as bees from hive, from mole-hill ants." Phineas Fletcher, again, has, " Earth a mole-hole, men but ants." Eliza Cook sees us as "ants in a mole-hill running in and out," and Mackay speaks of "the Babel-burrows," where "the little emmets come and go." So, too, in Watts—

> " Yet once a day drop down a gentle look
> On the great mole hill, and with pitying eye
> Survey the busy emmets round the heap,
> Crowding and bustling in a thousand forms
> Of strife and toil."

Watts, by the way, is particularly given to the ant-and-mole-hill idea—

> " Be gone for ever mortal things ;
> Thou mighty mole-hill, earth, farewell !
> Angels aspire on lofty wings,
> And leave the globe for ants to dwell."

These instances may, perhaps, be thought sufficient, but I cannot help also quoting country-gentleman Somerville, who, as being a country-gentleman, should, if any poet should, most assuredly have known the difference between ant-hills and mole-hills—

> " She looks beneath
> Contemptuous, and beholds from far this earth,
> This molehill earth and all its busy ants
> Lab'ring for life."

Some of my critics have said of me that I am microscopic in my fault-finding, and, now and again, that I do not make sufficient allowance for the license of poets. I had hoped that I had avoided these objections. However, two volumes of these trivial studies of the poets' natural history have been published, and nearly every chapter in each volume will be found to contain a well-founded complaint of the poets' neglect of some central fact of nature's teaching.

I am speaking, as I have always spoken, of the ante-Tennysonian epoch. Our latest Poet Laureate was one of the most accurately-scientific naturalists of the day, and how exquisite he is in his fidelity to fact I need not repeat.

For example, in my " Bird " volume, the poets cannot say too much about nightingales; yet nearly every one of them speaks of the glorious songster, not the allegorical Philomela, but the actual living bird before them, as *female* and as *English*—peculiarly, specially, English.

Or, for example, in the " Beast " volume : the fox is treated universally as a monster of cruelty, and, *therefore*, properly punished by being torn to pieces by hounds.

These are, each in its way, typical errors. The first makes nonsense and an absurdity out of one of Nature's prettiest and most powerful parables. The second presents the poets in a thoroughly unsympathetic and Philistine aspect.

This my third volume will be found to illustrate incidentally the same unnecessary and unpoetical blemishes—indeed, I need go no further for evidence of this than the present mole-hill-building ant. Why, I would ask, should any writers go out of their way to insist on ants making mole-hills ? Why not let them make ant-hills ? Such conceits may be " trivial," but when they occur in every poem the total is certainly not trivial. Error becomes then characteristic. Moreover, the inaccuracies are, as a rule, useless ones, and diminish the actual beauty of the verses in which they occur.

Tennyson is invariably correct, and what can be more beautiful than his similes, analogies, and metaphors taken from real life ?

In the present instance, apart from the absurdity of the poets in making ants making mole-hills, there is an illustration of my meaning as to the "uselessness" of most poetical mistakes. We find the ant busy on a mole-hill. Now, how does the mole-hill improve or adorn the fancy? Is the earth in any way whatever like a mole-hill? Is it even a good simile? Now it *is* like an ant-hill, for when viewed from the Olympus of genius it might seem to the imagination the agglomerated result of infinite individual labourers, each very small and weak, but by co-operation capable of considerable achievements. Though even then it is only strictly applicable to, say, a city, and if so applied forms a fairly accurate and delightful simile.

But a mole-hill! The heap of soil which a single subterranean labourer, working in the dark, throws up fortuitously and at random, and in a minute or two, and not only one, but a whole row of them, and never thinks any more of them, does not even know of their existence! It would, in fact, be very difficult indeed to think of any natural phenomenon that was less like the earth, less like the busy haunts of mankind, less sensible, less poetical, than a *mole*-hill. Yet with the word ant actually on their pens, on the paper, the poets one after the other talk rubbish about "us pismires" and "us emmets" toiling and moiling over a mole-hill.

That we human beings have much in common with the "innumerous emmet," has been most delightfully demonstrated by many writers from Homer to Lubbock, but philosophic Young is hardly felicitous when he speaks of the race as "vagrant emmets on an air-suspended ball." If there is one thing which ants are not, it is "vagrant." There are, of course, nomadic species, the Bedouins and Tartars of the formic kind. But the common ant is not a vagabond. As opposed to many gipsy insects, it is in Rommany language emphatically a "house-dweller." Its paths all lead into highways, and its highways all converge upon

a city. Nor are they fitly placed on an "air-suspended" ball.
For ants cannot remain upon a spherical surface. Their
foothold is curiously insecure, and they drop in the most
helpless way off places where most other insects walk
with ease and safety.

But there are many allusions, such as Shelley's—

> " The thronging thousands to a passing view
> Seemed like an anthill's citizens "—

which are both accurate and admirable. Thomson, lagging
behind, borrows the phrase of " this ant-hill Earth."

In its other poetical aspects the ant is " provident,"—" By
careful nature led, to make (in Clare) the most of summer's
plenteous day,"—and " industrious." It is always the "toiling " " lab'ring " ant :—

> " Turn on the prudent ant thy heedful eyes,
>   Observe her labours, sluggard, and be wise :
>   No stern commands, no monitory voice
>   Prescribes her duties or directs her choice,
>   Yet, timely provident, she hastes away
>   To snatch the blessings of the plenteous day.
>   When fruitful summer loads the teaming plain,
>   She crops the harvest and she stores the grain."

These lines of Johnson are, perhaps, the best upon a theme
which meets with such abundant and just appreciation, and
here and there I find passages that treat the insect's providence with some originality. Thus, more than one poet
admires the ant for providing itself in spring with "a house
against summer's heat," while others call the hard-toiling
neuter a " housewife," and address the sexless thing as " she."
Prior notes thus a special and very singular feature of the
creature's foresight :—

> " Tell me why the ant
>   In summer's plenty thinks of winter's want
>   By constant journey careful to prepare
>   Her store and bringing home the corny ear.

> By what instruction does she *bite the grain ?*
> Lest hid in earth and taking root again
> It might elude the foresight of her care."

> " The sage industrious ant, the wisest insect,
> And best economist of all the field :
> For when as yet the favourable sun
> Gives to the genial earth the enlivening ray,
> ——all her subterraneous avenues,
> And storm-proof cells, with management most meet,
> And unexampled housewif'ry she frames ;
> Then to the field she hies, and on her back
> Burden immense ! brings home the cumbrous corn :
> Then many a weary step, and many a strain,
> And many a grievous groan subdued, at length
> Up the huge hill she hardly heaves it home :
> Nor rests she here her providence, but nips
> With subtle teeth the grain, lest from her garner
> In mischievous fertility it steal,
> And back to daylight vegetate its way."—*Smart.*

Milton, as always, is characteristic :—

> " First crept
> The parsimonious emmet, provident,
> Of future, in small room large heart enclosed :
> Pattern of just equality perhaps
> Hereafter, joined in her popular tribes of commonalty."

King Solomon is responsible for many errors, and among others this of the providence of ants. Twice in his Proverbs he speaks of the insect laying up food in the summer—"having no guide, overseer, or ruler, provideth her meat in the summer and gathereth her food in the harvest." And it is said by naturalists, " devilling " for King Solomon, that the ants of Palestine do, as a matter of fact, hoard supplies for "the winter." Let this be as it may, the ants of Great Britain do nothing of the kind. These wonderful communities comprise, as is well known, males, females and neuters. Of these, the neuters alone do work in the sense which attracts the poet's admiration, and certainly if industry is ever praiseworthy, it is here. Their lives are one unceasing round of unselfish toil. The young, whether as eggs, grubs or pupæ, receive from them an amount of attention which even in a monthly nurse would be called unconscionably fussy. Every time the weather changes, the neuters shift the eggs and young so as to give them the advantage of as much warmth as possible, and when they are in their eating stages, are perpetually at their mouth with food. In winter, however, the community is dormant, torpid. The males and females are dead, and the surviving neuters hybernate in the lowest chambers of the nest, in company with their helpless charges. These facts are, therefore, damaging to the poetical pictures of the insect securely feasting in its barns during the winter months :—

> " The wiser emmet quoted just before
> In summer time ranges the fallows o'er,
> With pains and labour to lay in his store.
> But when the blustering north with ruffling blasts
> Saddens the year, and nature overcasts,
> The prudent insect hid in privacy,
> Enjoys the fruits of his past industry."

Nor is it correct to imagine that the race enjoys a longer life than other insects, for the males and females, which are the perfect insects, die, as all other insects do, within the twelve-month. A proportion of the neuters survive to the following spring, in order to tend the next year's brood, but not, so to speak, for the enjoyment of a renewed existence on their own account. Cowley, therefore, is not over fortunate in his moral—

> " Wisely the ant against poor winter hoards
> The stock which summer's wealth affords ;
> In grasshoppers that must at autumn die,
> How vain were such an industry !"

Nor Green in " The Fable of the Old Comedian," the grasshopper being the thriftless moralist of the poem—

> " My wretched end may warn green springing youth
> To use delights as toys that will deceive,

> And scorn the world, before the world them leave,
> For all world's trust is ruin without ruth.
> Then blest are they that, like the toiling ant
> Provide in time 'gainst woeful winter's want.
>
> With this the grasshopper, yielding to the
> Weather's extremity, died comfortless without remedy."

For the grasshopper lives quite as long as the ant, indeed longer, when we take the duration of the several stages of each insect's development into consideration.

Nor am I thoroughly in sympathy with those who uphold the ant as an exemplar of "wisdom." The male enjoys life as frivolously as any other insect of his sex. The female goes through her one duty of maternity with prodigious diligence. The neuter, poor little drudge, slaves out its twelvemonth of never-faltering labour with an apparent cheerfulness and unflagging alacrity, which is indeed amazing, bewildering.

But all human reason is ranged against this instinct for ceaseless moiling. If he could, St. Lubbock would, I am sure, insist on bank holidays for these unfortunate little drudges, and everybody who has ever written or spoken on the subject agrees that there is unwisdom in perpetual slaving. For myself, I have never greatly applauded the ant. It sets an example which, if man were to follow, would at once make life intolerable, and eventually result in putting lunatics in the majority. So I think the insect, as a pattern to the sluggard, should be suppressed, or at any rate that its abominable industry should cease to be quoted for our edification.

There is virtue in working. Nobody will deny that. But there is wisdom also in timely relaxation. I hope, therefore, it is true, as some observers say, that ants play at games. Meanwhile, I like Butler's lines :—

> " Those get the least that take the greatest pains
> But most of all i' the drudgery of brains,

> A nat'ral sign of weakness, as an ant
> Is more laborious than an elephant;
> And children are more busy at their play
> Than those that wisely pass their time away."

Lovelace, too, is delightful on the same theme. He has a poem on the ant, "Thou Miserable Ant," in which these verses occur :—

> " Austere and cynic, not one hour t' allow
> To lose with pleasure what thou gett'st with pain;
> But drive on sacred festivals the plough,
> Tearing highways with thy o'ercharged wain,
> Not all thy life time one poor minute live
> And thy o'erlaboured bulk with mirth relieve.
>
> Forbear thou great good husband, little ant,
> A little respite from thy flood of sweat:
> Thou thine own horse and cart, under this plant
> Thy spacious tent, fan thy prodigious heat,
> Down with thy double load of that one grain
> .   .   .   .
> Cease large example of wise thrift awhile
> (So thy example is become our law)
> And teach thy frowns a seasonable smile
> So Cato sometimes naked Florals saw."

In Fancy, the sober-sided, matter-of-fact, humdrum ant finds but small place. In the feast of the Fairies—when flies' eggs poached in moonshine, and butterflies' brains fricasseed in dew, a sucking mite roasted, and a rainbow tart, formed the bill of fare—the place of honour is given to the whole ant barbecued. A sprite riding on an ant is thrown, and "under the unruly beast's proud feet" lies mangled. It raises a monumental mound to the dead fay and befriends unhappy stepdaughters who are kind to it. Then there is the ant in amber :—

> " Whilst in an amber shade the ant doth rest,
> A gummy drop ensnares the small wild beast."

But Clare, referring, it may be, to the fairy origin of ants, which is current in many parts of the country, has a very delightful passage :—

> " What wonder strikes the curious, while he views
>   The black ant's city, by a rotten tree,
>   Or woodland bank !   In ignorance we muse :
>   Pausing annoyed—we know not what we see ;
>   Such government and thought there seem to be ;
>   Some looking on, and urging some to toil,
>   Dragging their loads of bent-stalks slavishly :
>   And what's more wonderful, when big loads foil
>   One ant or two to carry, quickly then
>   A swarm flash round to help their fellow-men.
>   Surely they speak a language whisperingly,
>   Too fine for us to hear : and sure their ways
>   Prove they have kings and laws, and that they be
>   *Deformèd remnants of the fairy-days.*"

CHAPTER II.

NIGHT-BUTTERFLIES AND DAY-MOTHS.

MOTHS are somehow supposed to be the poor relations of butterflies. When people who care little about such matters see a prettily coloured insect flying by, they say, " Look at that butterfly," but if it be dingy or insignificant in size they call it "a moth." Here, then, as everywhere else, the old fallacy "from authority" maintains its force. Holy Writ has damned the moth. So superstition and folklore approach it prejudiced. And Poetry, which used always to go to Holy Writ and Folklore for its natural history, falls into the ditch with its leaders.

But there is, of course, no ground for this invidious distinction between the butterfly and the moth. It is true that night-flying things, dusky things, do not commend themselves to general admiration with the same immediate attractiveness as those that delight in sunshine. The bat, being crepuscular—*entre chien et loup*—is a creature of shocking possibilities. Victor Hugo apostrophises it as a bird. Most poets call it "obscene." So, too, the owl, and the night-jar, and the "night raven"—that most delicious of all poetical bogies, invented by poets for their own titillation, just as children like to get themselves into the "creeps" by pretending an imaginary bear or giant under the bed.

They are all odious because they are of the twilight and the darkness, silent-flighted, mysterious in the gloaming.

Even so, too, is it with the moths. Scripture saith that it "corrupts." So it does—at any rate one moth, the "clothes moth." They *do* corrupt. As Bacon would say, they are "shrewd things" in a wardrobe—yet only think of the horrible "corruption" of which the good busy exemplary ant is guilty!

One iniquitous insect, however, has brought down upon the whole republic the everlasting condemnation of poetry. There is one tiny drab carpet-creeping creature of which the tiny grub, fearfully, eats cloth and clothes. Says Quarles—

> '' Like moths their houses shall they build, in doubt
> And danger, every hour, to be cast out.''

There are others just as inconsiderable, just as feeble in colouring, which plunder other possessions of our lordly race, apples and meal-bags for instance.

But what is there in all the range of insect beauty to excel our moths? What epithet appropriate to butterflies is excessive when applied to moths? For every '' beautiful,'' '' painted,'' '' gorgeous,'' '' spangled '' butterfly I will match you with a moth. More than this, I will bring you moths that shall defy all your butterflies to match for chastity of colouring, for exquisite design, and perfect beauty. Do you want audacious splendour of colouring or most delicately dainty tracery? Will you challenge me upon form or upon hue? I will meet you with such rare tints of all the colours as shall beggar you to surpass, and for form such elfin prettiness as shall be your despair.

It is a mistake, then, to think, as most do, that evening-flying moths are dull-coloured and dowdy, just as it is a mistake to suppose that those birds of wondrously beautiful plumage, the owls and night-jars, are unworthy of admiration. We have very few birds to equal them in the loveliness of their feathers and their colouring.

Nor are they small in size. Some of our English moths are larger than any of our butterflies. Nor, again, are they to be depreciated as night-flies, for many fly by day—and among them some of the prettiest. Moreover, in character, robust differences of temperament, and physical habit there is no comparison between the two. They are absolutely equal. Tell me of your purple emperor that tantalises you from the tops of oaks (and is caught eventually on a dead cat), and I will give you the sphinges—the humming-birds of our isles, the wondrous ones that make our petunia-beds tropical, our heliotropes and honeysuckles redolent of California and the Brazils. What splendour of wing! What a delicate adjustment of velocity! See them round the clematis: it is a spectacle to remember all your life. What mechanism of man's genius can equal the lightning-flash apparition? that instantaneous immobility, still as a star? that lightning-flash departure? If we could only imitate it, we should have sky-ships and sky-trains, and every man would call to his servant in the morning for his "wings," as he does now for his boots. Or take your stately flighted "white admiral," your luxurious fritillaries— Aglaia and the rest—or swift Edusa. Can I not match them?

The poets, therefore, knew nothing whatever of moths. They supposed that they "corrupt," because they knew that rust does, and that thieves break through and steal. But this is after all poor reasoning, or else sparrows ought to be selling at twelve a penny. Besides, translators of the Bible took extraordinary liberties with natural history. Our marginal readings warn us to read "badger" instead of "swan," and so on verse after verse. Albeit I consider the wondrous language of our translations as one of the most *positive* evidences of "inspiration" that has been vouchsafed to us.

Now, for myself, I have a singular sympathy with creatures

that are not of our own day; that make their own. Any pedestrian wretch can be abroad and awake in the sun-light. Only original genius—like Byron's, Shelley's, Goethe's, Lessing's, and everybody else's worth talking about—can resent the formalities of the clock and insist upon the right to choose its own hours. Nearly all the best work of the world has been done while the stupid were asleep.

That the moth is crepuscular; that in combination with rust it "doth corrupt"; that light has a fatal fascination for it, are among the ordinary facts of moth-lore that are common property, and constitute the sum of poets' science.

> "And every silver moth fresh from the grave,
>   Which is its cradle—ever from below
>   Aspiring like one who loves too fair, too far,
>   To be consumed within the purest glow
>   Of one serene and unapproachèd star,
>   As if it were a lamp of earthly light."—*Shelley.*

On the first account, they start with a general prejudice against it; on the second, they enjoy a specific reason for dislike to it; on the third, instead of pitying the dreadful fascination—surely one of the most horrible "instincts" in nature—they reproach the moth, sneer at it, and expatiate in common-place morals on youth and pleasure, beauty and desire, and so on. Here and there is a compassionate touch, a word of tender regret; but as a rule the analogies are neither compassionate nor tender, nor even excellent as imagery.

Shelley is an exception; he is one of our poets of the night, and the "silver moths" are therefore under his protection. There is originality in the line "Sweet Lamp! my moth-like muse has burnt its wings," and beauty in these—

> "Such clouds as flit
> Like splendour-winged moths about a taper
> Round the red west when the sun sets in it."

For the poets, for poetical reasons, are adverse to darkness and the things belonging thereto. Though constantly confessing to the practical advantages of night, they perpetually abuse it as a time of all dreadfulness. Yet we know it to be the tutelary genius of the small-creature world, their protection and Providence. Concealed by it, they repose in security, and under cover of the shadows of night come out to seek the food which they dare not look for when man and other large animals are abroad. Moths are among the illustrations of this protective caution.

Science, for the sake of classification, has butterflies, moths, and neither-one-thing-nor-the-others, the differences in the three being solely in the shape of the end of the antennæ. But practically there is no difference at any stage of life, either as caterpillar, chrysalis, or perfect insect. Nor in any habit or detail of life. Each division hoplessly overlaps the other. Moths are, therefore, simply night-butterflies. Butterflies are day-moths. Some fly by night, for better protection doubtless, but many by day. Thus among the heather in July, the "heath moth" and "heath butterfly" flutter about together. Poets, therefore, have no grounds for drawing a line between the two classes. But, as I have so often noticed before, the fact of any class of creature being *nocturnal* prejudices the poets against it. To be "dusky" is to be unamiable.

For instance, Thomson says—

> "Soft-buzzing Slander, silly moth, that eats
>   An honest name—"

as if moths "buzz!" as if there was any, the remotest, analogy between slander and a moth! Ben Jonson has—

> "Security—
> It is the common moth
> That eats our wits and arts, and so
> Destroys them both."

Change, as tending to waste, and Desuetude (the old

association with rust being in the poet's mind) are "moths."

But their treatment of these insects illustrates another large poetical obliquity. Because men and women who are gaily dressed are often frivolous and vain, it is assumed that gaiety of attire in the animal world leads to the same moral weakness.

As, for instance, among birds the goldfinch, among beasts the panther, among flowers the tulip, are reproached for being "gaudy," "gay," and "painted," and, *therefore*, "vain" and "fickle," so it is with the Butterfly. The poets concede its beauty, but, with that unfortunate tendency of their's to translate all animal life into human phrases, go on to assume that because it is fair to look upon therefore it is conceited, light in affection, heartless, proud.

> "Conscious of beauty they speed from flower to flower,
> Flaunting in the aspect of the day
> Their robes of spangled tissue."

Clare, a sympathetic poet, is one of the few who really gives them the advantage of a doubt.

> " For amid the sunny hour,
> When I've found thee on a flower
> (Searching with minutest gleg)
> Oft I've seen thy little leg
> Soft as glass o'er velvet glides
> Smoothen down thy silken sides ;
> Then thy wings would ope and shut,
> Then thou seemingly wouldst strut ;
> Was it nature, was it pride?
> Let the learned world decide."

It follows, then, naturally enough from the poet's moral, that "so gay a popelet" should be a fop, "light-fluttering," a "beau," and "dandy." But how ugly such metaphors are ! This, for instance—

> " And that thing was made of sound and show
>   Which mortals have misnamed a beau,
>   But in the language of the sky
>   Is called a two-legged butterfly."

Or this—

> " But for those butterflies the beaux
>   Who buzz around in tinsel rows,
>   Shake, shake them off with quick disdain.
>   Where insects settle they will stain."

I cannot understand how "poets" can even fancy so disagreeably, so sordidly, of these pretty harmless things with their happy ways—so happy, indeed, that they need no voice to tell us how glad they are of life, and how good they find the sunshine, and how fair the flowers. I think a better moral might easily be drawn from the flirtation of a butterfly and a daisy than this—

> " The dandy butterfly
>   All exquisitely dressed
> Before the daisy's eye,
>   Displays his painted vest,
> In vain is he arrayed
>   In all the gaudy show ;
> What business hath a maid
>   With such a foppish beau?"

I confess that, for myself, I am a trifle bored with the "innocence" of the daisy, and half wish some poet would get up and call it names. And I am quite sure butterflies do not go and show off before them, or even look as if they did. Nobody ever saw a butterfly on a daisy.

A worthless woman has this, again, for her epitaph—

> " Here lies, now a prey to insulting neglect,
>   What once was a butterfly gay in life's beam ;
>   Want only of wisdom denied her respect,
>   Want only of goodness denied her esteem."

Poor Psyche ! Who does not know the exquisite legend of Love and Psyche?—

> " Love's own,
> His wedded love—by holiest vow
> Pledged in Olympus, and made known
> To mortals by the type which now
> Hangs glittering on her sunny brow—
> That butterfly, mysterious trinket,
> Which means the soul (though few would think it)."

It is a "holiday-rover," "giddy," bent only on enjoyment, "to pleasure ever on the wing."

> " From flower to flower on balmy gales to fly
> Is all she has to do beneath the radiant sky."

So it is "the aimless butterfly" that flits "through ether without aim," "withouten any choice," or that, "fickle as a butterfly-love "—

> " doth rejoice
> Every minute to change choice,
> Counting he were then in bliss,
> If that each fair of all were his."

Moreover, singularly enough, the poets seem to resent the butterfly's elevation from the grub-commonalty to the Upper House. The moralists among them see in the metamorphosis an aspiring mind rewarded by success, or a pious one rising from earthly clods to heavenly heights. But, as a rule, it is remembered unkindly against the charming insect that it should have had such lowly origin, the intermediate refining stage of chrysalis-tranquillity notwithstanding. So, as the butterfly is born of the caterpillar, "to Goody Maggot near allied," and as those "reptiles," as poets call them, are (poetically) supposed to be engendered from corruption, the butterfly is a courtier-insect which is the outcome of a rotten society. In this character it perpetually recurs in verse. Again, inasmuch as the grubbing, toiling caterpillar may be imagined as having laid the foundation for the pleasures which the winged thing possesses, and which it so prodigally enjoys, it affords a simile for a spendthrift heir.

> " All glossy, gay-enamelled all with gold,
>   The silly tenant of the summer air
>   In folly lost, of nothing taking care."

Thus basely born, and being, as it were, fortuitous and ephemeral—not really butterflies, but only grubs on the wing—they are sometimes regarded as positively contemptible and beneath the appreciation of sensible men. Thus Shenstone, addressing collectors, says—

> " Hail, curious wights! to whom so fair
>   The form of mortal fly is,
>   Who deem those grubs beyond compare
>   Which common-sense despises."

So, too, Pope contemptuously relegates butterfly-collecting to "curious Germans." The insect is of no value—"swiche-talking is ne worth a boterflie."

But from all the above I do not mean it to be understood that the poets did not admire butterflies. On the contrary, their verse is full of compliments to their actual beauty. This admiration, however, is so very common, indeed so universal, that it would have been impossible for poets not to share in it; and quotation of their approving epithets is equally unnecessary. Indeed, it will be seen that even when depreciating the insects their physical charms are punctually admitted. None the less, it is curious to note how this wrong way of regarding the animal world—that is, from human standpoints—and transferring our own failings to all other creatures, is consistently adhered to by poets, who should remember that some writers, judging in the same erroneous way, have actually called the poets themselves " butterflies."

Some special phrases of admiration are, however, well worth noting; as for instance when the poet, *more suo*, exceeds the already-sufficing measure of the creatures' natural beauty. Thus Keats in fancy has a "golden" butterfly, but it was born of a flower, a fairy insect, and led

Endymion to strange and sweet adventure. But what real butterfly Shenstone had before him which he describes as "celestial crimson dropped with gold" it would puzzle the naturalist to say, or what the very common insect of poetry can be that has " wings of gold." "Gaudy, spangled wing," "painted and spangled," might apply to the fritillaries, but poets are too inaccurate as observers to have noticed the under side of these species' wings. Yet Darwin, who was a student of nature, has "silver" butterflies; while Wordsworth, who takes such credit to himself for communing in intimacy with nature, actually says he saw English boys chasing a "crimson " butterfly, and describes another "all green and gold "—two insects which it is perfectly certain he could never have beheld in Great Britain. Yet he tells us that he watched a single butterfly "a full half-hour," and that as a child—

> " My sister Emmeline and I
> Together chased the butterfly ; "

so that he ought to have known them by sight at any rate, and to have remembered that "sister Emmeline and I" never saw either a crimson or a green-and-gold butterfly.

They are "flying flowers," as the Chinese (or is it the Japanese ?) call them :—

> " One might fancy the rich flowers
> That round them in the sun lay sighing
> Had been by magic all set flying."

Summer comes on butterfly wings ; Pleasure is a butterfly : Youth "on its insect-wing " flies, " eager to taste the honey's spring." Keats beautifully calls them "lords of flowers." In Byron the butterfly affords an analogy for beauty wooed, won, and thrown aside :—

> " So beauty lures the full-grown child,
> With hue as bright and wing as wild,
> A chase of idle hopes and fears,
> Begun in folly, closed in tears.

> If won, to equal ills betray'd
> Woe waits the insect and the maid,
> A life of pain, the loss of peace,
> From infant's play, and man's caprice;
> The lovely toy so fiercely sought,
> Hath lost its charm by being caught,
> For every touch that woo'd its stay
> Hath brush'd its brightest hues away."

Several butterflies and moths are individually specified without any loss of beauty to the verse. Thus Hurdis' insect, "with saffron wing superb," that "zigzag dances o'er the flowery dell;" Clare's "plain-drest butterfly of russet dye," which—

> "As if awakened by the scythe's shrill sound
> Soon as the bent with ripeness 'gan to die,
> Was constant with the mower in the meadow-ground
> Flitting the with'ring swathe and unmown blossom round."

Or the other—

> "Lo! the arching heavenly bow
> Doth all his dyes on thee bestow,
> Crimson, blue, and watery green,
> Mixed with azure shade between,
> These are thine—thou first in place,
> Queen of all the insect race."

Which Jean Ingelow more pointedly particularises:—

> "Open velvet butterflies
> That swing and spread their peacock eyes
> As if they cared no more to rise
> From off their bed of camomile."

Crabbe is delightful:—

> "From the sweet bower, by nature form'd, arise
> Bright troops of virgin moths and fresh-born butterflies;
> Who broke that morning from their half-year's sleep
> To fly o'er flowers where they were wont to creep,
> Above the sovereign oak, a sovereign, skims
> The purple emp'ror, strong in wing and limbs,
> There fair Camilla takes her flight serene,
> Adonis blue, and Paphia silver queen;

> With every filmy fly from mead or bower,
> And hungry Sphinx who threads the honey'd flower
> She o'er the larkspur's bed, where sweets abound,
> Views ev'ry bell, and hums th' approving sound ;
> Poised on her busy plumes, with feelings nice
> She draws from every flower, nor tries a flow'ret twice."

Eliza Cook, whose butterflies are beyond all sober conception, has "blue" moths flit past in poem after poem, and to leave no room for doubt, she sometimes refers to the insect as "moth with azure wings." Moreover, her moths are the companions of the bee. Thus, "the moth is full dressed and the bee is about," "the golden moth and the shining bee Will seldom rest on the willow-tree," "aspen stalks move with the moth and the bee," and so on. But, I suppose, Eliza Cook ought never to be taken seriously. Other poets, however, see "happy moths sporting in the sunny summer's sheen," meaning, no doubt, "midges."

And what Darwin may mean by the lines—

> "Ye painted moths, your gold-eyed plumage furl,
> Bow your wide horns, your spiral trunks uncurl,"

I am at a loss to imagine. Clare was a nice observer, and the "ermine" and the "ghost" moths "that live in the silence and sweetness of night" met with frequent notice in his verse, flitting along the meadows, "climbing up the tall grasses"—

> "Dancing with lily-white wings o'er the dew,
> Perched on the down-beaded grass like a fairy."

"Tender-speckled" is Bloomfield's epithet for, no doubt, the "ermine." And Keats admirably speaks of a painted window as—

> "Diamonded with panes of quaint device,
> As are the tiger-moth's deep-damasked wings."

Many have "silver" moths—an epithet which is admissible, seeing that the white or very light-cloured sorts are the only

ones likely to have attracted the notice of such "impres-
sionist" persons as poets.

> "The silver moth enjoys the gloom
>   Glancing on tremulous wing through twilight bow'rs,
>   Now flits where warm nasturtiums glow,
>   Now quivers on the jasmine bough,
>   And sucks with spiral tongue the balm of sleeping flowers."

As a boy at school I was an inveterate "collector" of objects of natural history, and my "museum" was at once the envy and the despair of my contemporaries. That I neglected my duties for my pleasures goes without saying, as all boys do this. But I did so systematically, doggedly, and triumphantly. Fifteen years later I went back to my old school. They wanted to hear "an old boy's" adventures as a traveller and a war-correspondent, and I met many of my masters, and to them I confessed the unsuspected obliquities of my school career. I had never in seven years learned one single proposition of Euclid. No, not even the first. I had never achieved an algebraic equation. No, not one. I loathed mathematics with a profundity that even the loathing for rhubarb and magnesia could not surpass. So I shirked and cribbed during these lessons with such resolution and judgment that I was actually able to leave school congratulating myself on never having done a proposition or a sum in a legitimate way. At mathematical examinations I used to send up blank papers, or coolly return myself as absent.

In after life, when I first began to write for publication, I found myself handicapped by my total ignorance of precise terms, such as mathematics teaches, so, at my leisure, I read Logic and Metaphysics for three years, and thus qualified for the Professor's Chair of these sciences at the Allahabad College in India. The balance was thus equalised, you say, but was it? I found myself with all my sympathy for nature to the good. For no man, when his ideas are fixed, his character made, can deliberately, and in cold blood, set about putting himself in a business-like way in sympathy with nature. Such attempts end in mere sentimentalism, gush, and vulgarity. His love of nature must have been acquired when a boy. It can only come from association with wild things : the handling of flowers and birds' eggs, and butterflies ; the watching creatures going to and fro about their daily duties and pleasures—"idling," in fact, as some people call it, in the open air. Now, no grown-up man can methodically commence doing that. He has other things to attend to. But any adult can make up in three years, if he chooses, for the mathematical deficiencies of his teens.

Arithmetic has, by that time, sufficiently taught itself. Euclid turns out to be only common-sense. As for algebra, I have never been able to trace its influence anywhere—either in conversation, business, or literature. It *may* be very useful, just as writing lessons may be in "caligraphy," in which—O thrice-excellent Sellick, you will bear me out—I achieved as a boy considerable honours, my round-hand being (even to this day) a thing of beauty. And yet, gentle reader, I feel inclined sometimes to sit down and out of the fulness of my heart weep when I think of the kind of manuscript which I send in to printers. Those of my own household even cannot sometimes decipher my writing—if I may call it such. It is simply execrable. So that my writing-lessons have not permanently advantaged me any more than my evasive treatment of "Euclid" has injured me.

But there was something which I learned when at school which has never left me, and to which I am largely indebted for such measure of success as may have attended me, and that is "sympathy" with all sorts and conditions of creatures, founded upon a just appreciation of them—itself based on sound, first-hand, though assuredly very "unscientific" knowledge. I have learnt all my natural history from nature, who is the most unscientific teacher possible.

Of men and women I know little : their faces, gestures, demeanour are almost without significance to me. I am therefore always wrong in my estimates of men and women, and to my constant misfortune. For this reason I cannot read Dickens for more than a few pages at a time with pleasure. He bores me when he is not funny. For this reason I never—never—go to a theatre, except under compulsion, or else to laugh. On the other hand, I will not say how much I have read Darwin, or how often I go to the Zoo. My natural sympathies, therefore, are outside of humanity.

As for example : my able namesake F. W. Robinson the novelist would walk over Waterloo Bridge, and while walking over would receive a magazine-article-full of impressions —character sketches and so forth. I walk across with him, but on arriving at the other end I have simply traversed so many yards. That I passed a lot of people, I know, because most of them would not get out of my way—but little more. Well, perhaps a very pretty girl or an extraordinarily fat man may have awakened an instant's interest ; but virtually, after crossing the bridge, I find I have acquired no ideas whatever. My mind has received no impressions. Now, on the other hand, if I were to take F. W. across a meadow, he might go from stile to stile and at the end of the walk have nothing more than a general idea of grass, and, perhaps, a more definite one of sheep ; while I have reached the stile with enough in my head for an article in the *Gentleman's.* Put a magpie down on Waterloo Bridge, and the whole scene changes at once—*for me.* Put a shepherd and his boy into the meadow, and the whole scene changes at once —*for F. W.*

Now F. W., I admit (in deference to human perjudices), recognises and sympathises with a nobler range of subjects than I do. But admitting this, I still would not concede, even to him, the right of taking liberties with animals. If he should sketch a gluttonous, fat, lazy, sensual man, and end up by saying " in fact, a regular pig," the novelist would be just as far beyond his authority as if I, having described some animal as cunning, heartless, and voracious, should sum up with "in fact, a regular solicitor." The latter would sound the more unjust to the majority. Because every one knows that solicitors, taken as a class, are not necessarily cunning, heartless, or voracious ; and very many know, no doubt, individual solicitors who are positively straight-forward, kind-hearted, and moderate in fees ; while nobody, except a very few pig-breeders, knows anything of swine as a class, and it is only a small minority that can even speak to the character of any one individual porker. Yet the one generalisation would, in fact, be quite as unjust as the other.

How does this read? " The hyæna is as ugly as most old pawnbrokers, with the voice of a female politician, the shifty eye of a professional company-director, and a slink in the hind-quarters suggestive of a private detective. Though feeding as indiscriminately as a city alderman, it always, like a newspaper-editor, shirks conflict with those who can retaliate," and so on. How does it read? Does it not seem prejudiced and unsympathetic? That is just my complaint against the poets who, studying only mankind, insist on fitting in all the rest of creation into their ideas about humanity.

CHAPTER III.

ARACHNE AND THE POETS.

SPIDERS are best known by inference. Cobwebs; *ergo* spiders. But the insects [1] themselves are not much in evidence. It is not their nature to come to the front; they are of a conspiring kind.

Yet it is very curious, considering their extraordinary numbers, how comparatively seldom they are seen. A light brought suddenly into a dark room will sometimes betray one guiltily flat on the wall, with its legs all spread out round it in the very act of sudden surprisal. In cob-webbed cellars, too, there is at every turn the suspicion of long legs suddenly withdrawn into grey tunnels, of shrinkings away, and stealthy evanishments. But it is only the suspicion. During an autumn walk they may be seen in shrubberies or out in the country, basking complacently in the centre of their beautiful nets that a fly ought to be able to see a mile off—comfortable dowager-spiders, as Wyatt says, "obvious to flies." Or in summer, if you care to look in the garden, you may find them everywhere (and very beautiful many of them are), and you will see that most awful of little creatures, the cat-spider, hunting for prey on the woodwork of your window or the sunniest patches of the wall where the flies like to settle.

Laugh if you like, but in all seriousness it is very exciting

[1] I call spiders "insects" all through.—P. R.

to watch this queer little tiger at work. It moves hardly faster than the minute hand of a watch. Indeed it does not *seem* to move at all, but rather as if the fly it was stalking was a magnet drawing the spider towards it involuntarily. The eye can hardly note the legs stir at all; and yet, as you watch, the interval between the tiny beast of prey and its victim steadily, perceptibly, decreases. And at last it is within leaping distance. And it gets ready. But with what exasperating slowness! Will it *never* jump? And when it does, you do not see it. The act is too rapid for the eye to catch. But there all the same is the fly with the cat-spider on its back rolling over and over. And the eyesight of the small panther! The lynx is purblind by comparison. And as for its courage, the wild boar even does not excel it. Indeed, it is so brave as to be almost tame. If you threaten it with your finger, it turns fiercely on you, retreating backwards, and very soon it loses all fear of you, and will go after any fly it sees in spite of you.

In India I have, at Dholpur, seen the natives hunt ravine deer with the cheetah, and I have often gone hawking, both with the shahin off the wrist and the shikra in the hand. But the cat-spider is just as good, and if I had as much leisure as Queen Christina of Sweden, I should like to train these little insects.

Cruel? Not at all. At least not to my thinking—nor to that of Providence.

What a study in animal life it is, this crouching atom, so small and yet so fatally deadly. The ledge of the window-pane is a trifle rough; see how the spider takes advantage of every little pimple or speck of paint. Or it is smooth; see how it slips over the edge and, knowing it is quite out of its victim's sight, makes short rushes, coming cautiously up to peer over the edge after each rush to see if the fly has changed its position. And then when it gets exactly under the fly, watch it come up on the level! Is the fly washing its face ?   The spider advances.   Does the fly stop washing its face as if disquieted ?   The spider stops, too,   And then the lightning spring, the rough and tumble, the fearful tenacity of its ferocious grasp.   It is a wonderful bit of nature : straight from the jungles.   And educational ; giving glimpses under the surface ; a light on the real life-story of insects.

Moreover, this particular spider is curious among its kind in that it turns its head on its shoulders to look about it. If a bird flies past the window it turns like lightning.   It will watch a person cross a room.   Some of its eyes being on the very top of its head, it can see behind it.   No terrier ever looked more knowing or cocked its head more cleverly. All the same, it gives the little creature a very uncannily intelligent look.

This digression has taken me from my point—which was that spiders are better known by their works than by their persons.   In the poets this truth is singularly illustrated by the fact that fifty references to cobwebs might be found for every one to spiders.   If the insects were unsuitable for poets' purposes this would not be strange.   But, on the contrary, they are full of "morals," all of which are abundantly recognised in rhyme, and suggest to the fancy an almost unusual number of metaphors, similes, and images. The chief of these Southey compendiously reproduces in the following verses from his poem "To a Spider" :—

> " Weaver of snares, thou emblemest the ways
> Of Satan, sire of lies ;
> Hell's huge black spiders ; for mankind he lays
> His toils as thou for flies.
> When Betty's busy eye runs round the room,
> Woe to that nice geometry, if seen !
> But where is he whose broom
> The earth shall clean ?
>
> Spider ! of old thy flimsy webs were thought,
> And 'twas a likeness true,

> To emblem laws in which the weak are caught,
>   But which the strong break through.
> And if a victim in thy toils is ta'en,
>   Like some poor client is that wretched fly—
> I'll warrant thee thou'lt drain
>   His life-blood dry.
>
> And is not thy weak work like human schemes
>   And care on earth employed !
> Such are young hopes and love's delightful dreams
>   So easily destroyed !
> So does the statesman, whilst the avengers sleep,
>   Self-deem'd secure, his wiles in secret lay,
> Soon shall destruction sweep
>   His work away.
>
> Thou busy labourer ! one resemblance more
>   Shall yet the verse prolong,
> For, spider, thou art like the poet poor,
>   Whom thou hast helped in song.
> Both busily, our needful food to win,
>   We work, as nature taught, with ceaseless pains—
> Thy bowels thou dost spin,
>   I spin my brains."

Parallel passages to each of these may be easily quoted—for Southey was a careful reader and a generous borrower. Thus, as to the web-weaving of Beelzebub, from Burns :—

> " Ah Nick ! ah Nick ! it is na fair,
>   First showing us the tempting ware,
>   Bright wines and bonnie lasses rare,
>   To put us daft,
>   Syne weave, unseen, thy spider's snare
>   O' hell's damned waft."

The simile of law, solicitors' clerks, "human insects catering for human spiders "—"men of law " (Crabbe) :—

> " There in his web th' observant spider lies,
>   And peers about for fat intruding flies.
>   Doubtful at first, he hears the distant hum,
>   And feels them fluttering as they nearer come ;
>   They buzz and blink and doubtfully they tread
>   On the strong bird-lime of the utmost thread ;

> But when they're once entangled by the gin,
> With what an eager clasp he draws them in.
> Nor shall they 'scape till after long delay,
> And all that sweetens life is drawn away."

And then as to the brittleness of the laws (Beattie):—

> " Laws, as we read in ancient sages,
> Have been like cobwebs in all ages—
> Cobwebs for little flies are spread,
> And laws for little folks are made ;
> But if an insect of renown,
> Hornet or beetle, wasp or drone,
> Be caught in quest of sport or plunder,
> The flimsy fetter flies in sunder."

The statesman's futile subtlety (Garth) :—

> " Or spider-like, spin out our precious all,
> Our more than vitals spin (if no regard
> To great futurity) in curious webs
> Of subtle thought and exquisite design
> (Fine network of the brain) to catch a fly !
> The momentary buzz of vain renown,
> A name ! a mortal immortality."

Or Churchill's—

> " Henceforth, secure, let ambushed statesmen lie,
> Spread the court-web, and catch the patriot fly."

And hopes in general (Wordsworth) :—

> " Hopes, what are they?   Beads of morning
> Strung on slender blades of grass ;
> Or a spider's web adorning
> In a strait and treacherous pass."

The simile of the poor poet spinning his brains into "lines" is used more than once, and also applied to other "lines" of thought, as in Cowper, of commentators :—

> " Comment after comment, spun as fine
> As bloated spiders draw the flimsy line,"

and Churchill's critics :—

> " Let wits, like spiders, from the tortured brain
> Fine draw the critic web, with curious pain."

When addressed as the actual insect in nature, spiders are generally "cruel," "delusive," "false," and "venomous." One poet calls them "villains;" another, "wily ruffians, gaunt and grim." They are "blood-bloated" and "all-bellied." Yet it is immensely to the credit of poets that they should in the case of this insect—so notoriously unpopular and unprepossessing—have been so often fair to it. Such justice is eminently unpoetical—that is to say, extremely rare in poetry, for, as a rule, poets adopt vulgar prejudices, and punctually inculcate them. It is true that in legends spiders are more often benign than malignant, affording another illustration of the tendency of folk-lore to look upon ugliness only as the masquerading domino of an enchanted goodness, and on misfortune in figure as the disguise of fortune of other kinds. So toads are often beneficent; serpents nearly always so. Humanity, in the real heart of it, is tenderly sympathetic. When the proper time comes everybody hopes bad will turn back to good, and ugly to beautiful. Folk-tale always ends in the recovery of lost possessions, the resumption of original charms.

But poets are not, as a rule, appreciative of the deeper sense of animal legends. Superficial ideas, such as the accidents of Holy Writ or the traditions of heraldry afford, they reproduce and elaborate, but they seldom catch the true spirit of the humanities about animals. So it has always seemed to me. The present instance, then, is exceptional and of some interest.

Its explanation, no doubt, is the poets' admiration of industry. So the spider is "grave," "patient," "industrious," and "a house-wife," and more than one bard comes forward to directly praise the insect. Southey, a very hard-working man himself, has a strong fellow-feeling with it :—

> "Spider ! thou need'st not run in fear about
> To shun my curious eyes,

> I won't humanely crush thy bowels out
> Lest thou shouldst eat the flies ;
> Nor will I roast [1] thee, with a damn'd delight
> Thy strange instinctive fortitude to see ;
> For there is one who might
> One day roast me.
>
> Then shrink not, old free-mason, from my view,
> But quietly, like me, spin out the line ;
> Do thou thy work pursue
> As I will mine."

Mackay has an excellent defence to offer in the poem commencing—

> " Though fear'd by many, scorned by all,
> Poor spider on my garden wall ;
> Accursed as ugly, cruel, sly,
> And seen with an averted eye,
> Thou shalt not lack one friend to claim
> Some merit for thy injured name."

So he takes each charge in turn.   As for its being "ugly," says he, you have only to look closer to see "a creature robed in brilliancy, with supple and resplendent limbs" and as for "cruel!"—must not spiders eat to live?   And if thou, poor spider, art cruel—

> " Because thou takest, now and then,
>   A fly, thy mutton, to thy den,"

what shall we call man who is perpetually killing, not from necessity, but for amusement?

> " And then we call thee sly, forsooth,
>   As if from earliest dawn of youth
>   We did not lay our artful snares
>   For rabbits, woodcocks, larks, and hares ;
>   Or lurk all day by running brooks
>   To capture fish with cruel hooks,
>   And with a patient deep deceit,
>   Betray them with a counterfeit."

---

[1] Southey, in his rummaging of zoological legends and folk-lore, was familiar with the frequency with which roasting spiders as specifics against various evils is inculcated.—P. R.

And then he goes on to plead that the beauty of the spider's skilful web alone "might for some venial faults atone;" while its patience in calamity, "its courage to endure or wait," its "self-reliance," claim several stanzas :—

> "Should stormy wind or thunder-shower
>   Assail thy web; in evil hour
>   Should ruthless hand of lynx-eyed boy,
>   Or the prim gardener's rake, destroy
>   The clever mathematic maze
>   Thou spreadest in our garden ways,
>   No vain repinings mar thy rest,
>   No idle sorrows fill thy breast.
>
> Thou mayst perchance deplore thy lot,
>   Or sigh that fortune love thee not ;
>   But never dost thou sulk and mope,
>   Or lie and groan, forgetting hope ;
>   Still, with a patience calm and true,
>   Thou workest all thy work anew,
>   As if thou felt that Heaven is just
>   To very creature of the dust.
>
> And that the Providence, whose plan
>   Gives life to spiders as to man,
>   Will ne'er accord its aid divine
>   To those who lazily repine."

Historical spiders—except that which, with the assistance of the "bird of Mecca," saved Mahomet from his pursuers, and the other that Bruce said he took inspiration from, are not honoured in verse. The latter legend is very befittingly done into rhyme by Eliza Cook. What does the patriotic Scot say to this as a setting for that immortal myth :—

> "It soon began to cling and crawl
>   Straight up, with strong endeavour ;
>   But down it came with a slippery sprawl,
>   As near to the ground as ever."

Nor may the spider of the dungeon in Chillon be forgotten :—

> "With spiders I had friendship made,
>   And watch'd them in their sullen trade,
>   Had seen the mice by moonlight play,
>   And why should I feel less than they?

> We were all inmates of one place—
> And I, the monarch of each race,
> Had power to kill—yet, strange to tell !
> In quiet we had learn'd to dwell."—*Byron.*

Of legends of the insect several are worth notice. Thus, that which Phillips, in his "Cider"[1] refers to :—

> "Happy Ierne ! whose most wholesome air
> Poisons envenomed spiders, and forbids the baleful toad ; "

and Green notices in the line,

> "As spiders Irish wainscot flee."

The superstition being that, not only did St. Patrick drive all "vermin" out of the island, but communicated to bog-oak the property of keeping spiders off. Is it not a tradition that there are no spiders, for this reason, in the House of Commons, the woodwork being all of Irish oak ?

"That Apulian spider's poisonous sting, healed by the pleasing antidote of sonnets," is, of course, the tarantula, as specified in Oldham :—

> "Some are at sound of christened bell forgiven,
> And some by squirt of holy water shriven ;
> Others by anthems played are charmed away,
> As men cure bites of the tarantula."

Or in Herbert's lines :—

> "As peculiar notes and strains
> Cure tarantula's raging pains."

That the spider is venomous is of course a scientific fact. Its jaws, so to speak, are perforated exactly like a viper's tooth, and communicate with a poisonous secretion, which thus passes into any wound inflicted. In the poets the fatal gift is of course as much exaggerated as the deadliness of

---

[1] The context is as follows :—

> "More happy in her balmy draughts (enriched
> With miscellaneous spices) and the root
> For thirst-abating sweetness praised, which wide
> Extend her fame."

What plant is the poet referring to?

owls' hootings or the balefulness of toads' spittings.   But
never surely so delightfully set forth as in Spenser's original
exposition of the Vanities :—

> "An hideous dragon, dreadfull to behold,
>   Whose backe was arm'd against the dint of speare,
>   With shields of brasse that shone like burnisht golde,
>   And forked sting, that death in it did beare,
>   Strove with a spider, his unequall peare ;
>   And bad defiance to his enemie,
>   The subtill vermin, creeping closely neare,
>   Did in his drinke shed poyson privilie ;
>   Which, through his intrailes spredding diversly,
>   Made him to swell, that nigh his bowels burst,
>   And him enforst to yield the victorie,
>   What did so much in his owne greatnesse trust.
>   O, how great vainnesse is it then to scorne
>   The weake, that hath the strong so oft forlorne !"

And King Endymion had a cohort of gigantic spiders,
and they spun a web from the Moon to the Morning Star,
and upon this, the Field of Gossamer, was fought the most
fantastic fight that ever the gods saw.   Three days it raged,
the Moon-folk and the Sun-folk fighting; till King
Endymion cut away the web, and thus, in one appalling,
overwhelming act, closed the fearsome strife.   Do you
remember, too, how in Laputa the professor intended that
spiders should supersede silkworms, for, said he, they not
only spin silk but they weave it as well; and how to make
the tissue complete, he fed them on beautifully tinctured
flies?   And why should he not have done so ?—

> "I've plans for weaving velvet
>   From the spider's web so thin."

In the Islands of the Blessed they are apparelled in spiders'
webs, naturally stained purple, for Idmon, the dyer in purple,
was father of Arachne.

> "Weaving spiders come not near ;
>   Hence, you long-legged spinners, hence !"

Poor spiders !   Titania wore their web, and yet would not
have her weavers near her.

"Hence, you long-legged spinners, hence!" Yet Puck swings on their threads, "hung between two branches of a briar," the fairies' lutes are strung with gossamer; their steeds reined with "smallest spiders' web," and Titania's "waggon-spokes made of long spinners' legs."

. . . . . .

"Then the Interpreter showed them into the very best room of the house, a very bare room it was, so he bid them look round about, and see if they could find anything profitable there. Then they looked round and round—for there was nothing to be seen but a very great spider on the wall—and that they overlooked.

"*Mercy*. Then said Mercy, Sir, I see nothing.

" But Christiana held her peace.

"*Interpreter*. But, said the Interpreter, look again.

"She, therefore, looked again, and said, Here is not anything but an ugly spider, who hangs by her hands upon the wall.

"Then said he, Is there but one spider in all this spacious room ?

"Then the water stood in Christiana's eyes, for she was a woman of quick apprehension ; and she said, Yea, Lord, there *are* more here than one. Yea, and spiders whose venom is far more destructive than that which is in here.",

This is a delightful passage, spoilt, perhaps, to some by the fact that the translators of the Bible ought probably to have said lizards instead of spiders, and that the "semam" of the original is still an undetermined quantity. Not that such a detail disturbs the poets. For them this insect is of one species only and without varieties. It is "the spider"—that spins web "out of her own bowels." And yet how prodigious, far-reaching, and deep-searching in its influences on the economy of insect life is this many-tribed insect.

. . . . . .

Who has not at one time or another thought the lives of

insects enviably free from care? When out among the heather, or resting some summer's noon under a tree, how happy the small-winged folk, sunning themselves on the flower-heads, seem to be, without, apparently, any troubles or even responsibilities. They zigzag and flutter about as if time and place were nothing to them. This field or the next—what does it matter? Now, or by-and-by?

But whenever I hear any one envy the life of flies, I think to myself, "My friend, you have forgotten *the spiders.*" Sunshine and wings and flowers—a delightful combination, no doubt. But think of the spiders. It is worth while to do so, for it may turn over quite a new page in nature for you.

Imagine, for instance, that the insects which you see "dancing" in the air, out of pure light-heartedness, too happy even to sit still on a flower, or even to make their choice on which one they will settle, as you imagine, *are afraid to alight.*

Fancy, if you can, that every blossom, every tempting twig, has a hungry spider upon it, and that the flies know it, and dare not rest. What, then, becomes of all their light-heartedness, of the gladsomeness that keeps them so buoyantly ever on the wing? From every resting-place, bright petal or green leaf, keen, patient eyes are looking out and up at the winged things half-minded to settle and yet afraid. Fangs are working and mumbling together in the excitement of expectation, legs are drawn up all ready to spring—and the fly knows it. It hovers over the welcome perch, the tempting honey, but instinct tells it of a peril that is ambushed; its courage fails it, and, just as you think it is going to settle, it is gone. How "frivolous," "giddy," and all the rest of it, it seems, this fly in the summer's sunshine, dancing from flower to flower, does it not? But, are you sure that the fly did not see something? Look close yourself at the purple cushion of that scabious. Nothing? Look closer. Nothing still? Look underneath. Ah? Now, suppose yourself a fly, and that spider as large as yourself, and then conceive, if you can, the blood-curdling horror of such an apparition suddenly confronting you. If you had human wits about you before you met it, the odds are that you would be a gibbering idiot for ever afterwards. Human reason could not possibly stand the shock of such a fearful sight. Spiders the size of bullocks would kill at sight.

> " Mock the majesty of man's high birth,
>   Despise his bulwarks, and unpeople earth."

No wonder then that the fly was reluctant to settle on the scabious, and that it " danced " about the flower so long, and eventually decided not to sit down. But the "giddiness" and "light-heartedness" has all gone out of the picture.

Of course I do not assert that my idea is correct. It may be only a fancy. But it is, at any rate, perfectly safe to assume that, in a very large number of cases, the precipitancy of a fly's departure is due to a very proper discretion, and not to silliness. Also, that very often indeed when an insect seems unreasonable in its sudden changes of intention, it has, as a matter of fact, the best of all reasons for its conduct, namely, escape from death.

The purple cushion of the scabious, so warm with the sun shining full on it, and each of the little flowerets that compose the disc so full of fragrant honey, is the very ideal of a resting-place for a fly. And so, too, thinks the fly, till there grows up gradually over the edge of the flower two fine green legs tipped with little claws. Then it is time to be off—there is none to waste. If the fly stands upon the order of its going, there will follow the legs a pair of grass-green nippers, exquisitely sharp at the points, toothed, too, on the inner side, and hollowed like a cobra's fang to carry poison. And above the grass-green nippers will be two rows of eyes as bright as diamonds—and that is the last the fly will remember. So it wisely goes at once.

The knowledge of this prevalence of spiders goes some way to give a just appreciation of insect life. At any rate, without it, any idea of that life must be as ridiculously incomplete as if we were to think of mice in kitchens without cats and traps.

For myself, I know this perpetual recurrence of spiders and their webs in every crook and cranny of the country makes me think of insect existence as anything but careless. It rather seems to me a constant ambuscade. Fortunately, the flies do not know it, or I should think they would abondon life in despair; and fortunately, too, they have no nerves, or they would go mad with horrors and the apprehension of them. To me, then, there is always present in the quietest scenes in nature an underlying grimness which makes insect life very real and serious.

The birds seem merely an awkward incident in fly-life; their prodigious havoc among the winged things only occasional and local as compared with the ubiquitous and universal spiders'. The former, no doubt, are active and voracious and very numerous. But what is their activity, voracity, or their numbers by comparison with the myriads of these little, subtle, and nimble blood-suckers? They are everywhere—among the grass, in the herbage, the undergrowth, the shrubs, the trees, in the hedges and the ditches, on railings, tree-trunks, walls, and the ground; creeping, hiding, web-spinning, leaping, they cover the earth in a universal conspiracy.

Darwin's revelation of that awful force, the earth-worm, revolutionised my ideas of the under-world. Till then I had considered it inert or passive, the solid basis and ground-work of the shifting, active life of the upper-world. I had never regarded it as itself a scene of a pitiless diligence in ruin, the laboratory of a vast disintegrating agency that is incessantly toiling to bury cities and destroy every vestige of man's occupation of the earth.

With something of the same completeness came upon me the discovery of the all-pervading presence of spiders. Science calls them by the names of beasts of prey, and it was well to do so. For if you will take a foot of ground out in the country any summer's day as your sphere of observation, and watch for a while, you will see the cat-spider come creeping along, suddenly springing as it goes at everything that looks like a fly; the wolf-spider pass rapidly across with business-like directness; the lynx-spider sidle from blade to blade. Or spread a handkerchief under a bush, and strike the branches with a stick. Spiders come tumbling out, or hang in mid-air by the threads that, even against so sudden an alarm, they have all prepared.

So far, then, the spider in verse. But, as I said at starting, the creature is infinitely better known by its handicraft than its presence, and poetical references to its web are in porportion more numerous. There is, however, but little variety in the treatment. Occasionally the web is called "arras," "tapestry," "the lonely spider's thin grey pall," a "clue," a "bower," but, as a rule, it is either the open air "filmy," "silken" thread that catches the garden fly :—

> "So the false spider when her nets are spread,
> Deep ambush'd in her silent den does lie ;
> And feels far off the trembling of her thread,
> Whose filmy cord should bind the struggling fly,
> Then if at last she finds him fast beset,
> She issues forth and runs along her bower ;
> She joys to touch the captive in her net
> And drag the little wretch in triumph home."—*Dryden*.

Or the dusty-covered cobweb of the attic or cellar and neglected library :—

> "For a deep dust (which time does softly shed
> Where only time does come) their covers bear,
> On which grave spiders streets of web had spread
> Subtle and slight, as the grave writers were."

Or other vacant rooms where—

> " Her disembowelled web
>   Arachne in a hall or kitchen spreads
>   Obvious to vagrant flies ; ' she secret stands
>   Within her woven cell ; the humming prey,
>   Regardless of their fate, rush on the toils
>   Inextricable ; nor will aught avail—
>   Their arts, or arms, or shapes of lovely hue.
>   The wasp insidious, and the buzzing drone,
>   A butterfly, proud of expanded wings
>   Distinct with gold, entangled in her snares
>   Useless resistance make : with eager strides
>   She, towering, flies to her extended spoils."—*Wyatt.*

Or again in Montgomery :—

> " Around thy bell o'er mildewed leaves
>   His ample web a spider weaves ;
>   A wily ruffian gaunt and grim,
>   His labyrinthine toils he spreads
>   Pensile and light, their glossy threads
>   Bestrewed with many wing and limb ;
>   Even in thy chalice he prepares
>   His deadly poison and delusive snares.
>
>   Swift as death's own arrows dart,
>   On him the spider springs,
>   Wounds his side—with dexterous art
>   Winds the web about his wings ;
>   Quick as he came, recoiling then,
>   The villain van'shes in his den.
>
>   The desperate fly perceives too late
>   The hastening crisis of his fate ;
>   Disaster crowds upon disaster,
>   And every struggle to get free
>   Snaps the hopes of liberty,
>   And draws the knots of bondage faster."

This " triumphant " descent of the spider upon its victim is a very favourite fancy of the poets ; and, though I have never myself recognised any such exultation in the demeanour of the little fly-catcher, I have often imagined that I detected a high-stepping affected way about it when, having done up its prey into a parcel, it minces off to its

dining-room with its dinner dangling behind it.   At any rate,
there was abundant self-satisfaction, and not without cause
either.   For, taking nature all round, I know no episode
that excels in interest the successful web-spinning of spiders.
Savages are laboriously ingenious in trap-setting, but
"Arachne" gives them points at every stage.   And how
curious the poets' fancy of nature admiring herself, so
satisfied with the perfection of her own spider that she
copies it in flowers :—

> "Fair Cypripedia, with successful guile,
> Knits her smooth brow, extinguishes her smile—
> A spider's bloated paunch and jointed arms
> Hide her fair form, and mask her blushing charms ;
> In ambush sly the mimic warrior lies,
> And on quick wing the panting plunderer flies."—*Darwin.*

Did the poets know that "gossamer" was spiders' web?
Many certainly did not ; some are doubtful : some undoubtedly did.   Thus, Darwin, accurately excellent, speaks of the
adventurous flight of the newly hatched gossamer-spider :—

> "So shoot the spider brood at breezy dawn
> Their glittering network o'er th' autumnal lawn ;
> From blade to blade connect with cordage fine
> The unbending grass, and "live along the line."

And again, in "Prince Arthur," is the line :—

> "On the buoyant air sublimely borne."

And again, in Charlotte Smith :—

> "Small, viewless aeronaut, that by the line
> Of gossamer suspended, in mid-air
> Float'st on a sunbeam.   Living atom, where
> Ends thy breeze-guided voyage?   With what design
> In æther dost thou launch thy form minute,
> Mocking the eye?   Alas, before the veil
> Of dense clouds shall hide thee, the pursuit
> Of the keen swift may end thy fairy sail."

And here too ever-welcome Hardis :—

> "So dangles o'er the brook, depending low
>   The spider artist, till propitious breeze
>   Buoy him athwart the stream : from shore to shore
>   He fastens his horizontal thread,
>   Sufficient bridge, and, traversing alert,
>   His fine spun radii flings from side to side,
>   Shapes his concentric circles without art,
>   And, all accomplish'd, couches in the midst,
>   Himself the centre of his flimsy toils."

But some were in doubt (and I even suspect Southey), while others undeniably held the theory that the "gossamer" was condensed dew. That Spenser should speak of "the fine nets of scorched dew" is not remarkable, nor that—is it Cowley?—should have :—

> " Cobwebs that do fly
>   In the blue air, caused by the autumnal sun
>   That boils the dew that on the earth doth lie."

Nor, perhaps, that Quarles should say—

> " And now autumnal dews are seen
>   To cobweb every green."

But when I find Thomson speaking of "the filmy threads of dew evaporate," the superstition seems to me to have lived too long, even among poets.

As "gossamer" the poets perpetually admire the glistening threads. But as cobwebs they abhor it, especially (which I think delightfully poetic) when they remember it is spun out of the insect's "bowels." Sometimes, as in Pope's appreciative lines on "the spider's parallel designs," or Crabbe's straightforward admiration of its diligent "geometry," Arachne fares well at her wheel, but, as a rule, she is considered sinister and treacherous :—

> " The subtle spider never spins
>   But on dark days her shiny gins ; "

and her devices for securing food, so patiently worked out, so admirably efficient when complete, are looked upon as wicked frauds upon the confiding flies, snares for the inno-

cently unwary.   Yet it should be noted, as Butler notes it,
that—

> " Spiders never seek the fly,
> But leave him of himself to apply."

Herein the justice of Dryden's wit, "obvious to vagrant flies."
In metaphor they are usurers, attorneys, murderers, and
Jews.   But, for myself, greatly as they mar my pleasure in
the country by constantly intruding upon pleasant scenes
the evidences of strife and suffering, I have somehow as kind
a liking for spiders as for flies, and, as Scott says, feel as
tenderly for spiders "as if I were a kinsman of King Robert
of happy memory."   Nor would any notice of the poets'
spider be complete without quotation of Spenser's admirable
stanzas :—

> " In this gardin, where young Clarion
>   was wont to solace him, a wicked wight,
>   the foe of fayre things, th' author of Confusion
>   the shame of Nature, the bondslave of Spight
>   had lately built his hateful mansion
>   and, lurking closely, in awaite now lay
>   how he might any in his trap betray.
>
> And weaving straight a net with manie a fold
>   about the Cave, in which he lurking dwelt,
>   with fine small cords about it stretched wide
>   So finely sponne that scarce they could be spide.
>
> Not anie damzell, which her vaunteth most
>   in skillfull knitting of soft silken twyne,
>   nor anie weaver, which his worke doth boast
>   in diaper, in damaske, or in lyne,
>   nor anie skil'd in workmanship emboss'd
>   might in their diverse cunning ever dare
>   with this so curious networke to compare.
>
> Eftsoones her white streight legs were altered
>   to crooked crawling shankes, of marrowe empted ;
>   and her fair face to foule and loathsome hane
>   and her fine corpes to a hag of venim graine.
>
> Eftsoones that damzell, by her heavenly might
>   she turned into a wicked butterflie,
>   in the wide aire to make her wand'ring flight ;
>   and all those flowres with which so plenteouslie

her lap she filled had, that bred her spight
she placed in her wings, for memorie
of her pretended crime, though crime none were ;
since which that I lie them in her wings doth beare.

His breast plate first that was of substance pure
before his noble heart he firmely bound,
that mought his life from yron death assure,
and ward his gentle corps from cruell wound,
for it by arte was framed to endure
no lesse than that which vulcan mad e shield,
Achilles' life from fate of Troyan field.

And then about his shoulders broad he threw
an hairie hide of some wilde beast, whom he
in salvage forest by adventure slew,
and reft the spoyle, his ornament to bee
which, spredding all his backe with dreadfull view
made all, that him so horrible did see
thinke him Alcides with the lyon's skin,
when the Nœmean conquest he did win.

Upon his head his glittering burgamet,
the which was wrought by wonderous device
and curiously engravin, he did set,
the metal was of rare and passing price ;
Not Bilbo steele, nor brass from Corinth fet,
Nor costly Aricalchi from strange Phœnicet,
But such as could both Phœbus' arrowes ward,
and th' hayling darts of heaven beating hard.

Therein two deadly weapons fix'd he bore
strongly outlanced towards either side,
like two sharp speares, his enemies to gore
Like as a warlike brigandine, applyde
to fight, layes forth her tread full pike afore,
the engines which in them sad death doo-bayde ;
so did this flie outstretch his dreadful hornes,
yet so as him their terror more adornes.

Lastly, his shinie wings, as silver bright
painted with thousand colours, passing farre
all painters' skill, he did about him dight :
not half so manie sundrie colours arre
in Iris' bowe, ne heaven doth shine so bright
distinguished with manie a twinckling starre
nor Junoe's bird, in her ey-spotted traine
so many goodly colours doth containe."

CHAPTER IV.

FLIES: "THE HOSTS OF ACHOR."

FLIES and spiders are naturally associated. To the fly the
association is grim enough, and among the most appealing
sounds in nature is the long-drawn cry of the captive shrilling
its life out in the spider's web. And yet, extending a sufficient sympathy to the victims, I am tolerably content to
know that the spinner's toil has not been all in vain.
When I see the glistening husk of an empty bluebottle
hanging outside a cob-webbed chink, rattling like a calabash
on a savage's door-post, I drop the tributary tear to departed
worth ; but remembering, anon, the burly trespasser that
one slumberous afternoon in midsummer filled my room
for an hour together with the intermittent horror of its buzz—
as maddening in the suddenness of its cessations as in the
heartiness of its recommencements, now quartering the
carpet, now hurtling about on the ceiling, everywhere by
turns and nowhere long—I cannot but admire the kindliness
of nature in bestowing on the poor spider the gift of spinning
webs.

> " Wouldst fill the wholesome universe with flies
> and make the air too thick for human health ?
> Death is no evil. Cease, O foolish man
> thy querulous moaning, and consider death
> no longer as thy foe."

Those who think that the foremost function of the poet
is the drawing of morals should be content with the treatment which "flies" receive in verse.

At the outset, however, it should not be forgotten as I have carefully proved in a previous page that, poetically, insect, vermin, and reptile are interchangeable terms. Thus despicable humanity in general; or specific classes of unpopular persons and unpopular callings; or individuals held in contempt—are each and all described indifferently as insect, vermin, reptile. But the first of these terms is used to denote the "winged" variety of the obnoxious. Not as an invariable rule, of course, for poets do not hesitate to speak of both vermin and reptiles as "by the light air upborne." Still, as a broad distinction, insects mean in poetry *winged* insects, and still more specifically "flies." And though among these occasionally the butterfly, moth, dragon-fly, beetle, and other creatures are spoken of as flies, that word may be accepted as individualising those household insects, the domestic *musca*, and the bluebottle. Vanities and the transient and fleeting phases of human life are, of course, symbolised under the ephemerids—"the swarm that in the noontide beam are born." The May-fly, fire-fly, horse-fly, and so forth, are each used for their special purpose. But in the first place, and the majority of places, the fly of poetry is the common object of our window panes, the *imagines Diavoli et hereticorum* of Luther—in fact, the house-flies.

So far by way of premise. Now the poet's vermin and reptiles are, we have seen, "engendered" by the sun's heat out of mud and slime. In this origin "insects" therefore share. "Sire of insects, mighty Sol!" says Pryor, apostrophising flies. It is "the rank fly," "corruption's insect," Pope's

> " Morning insects that in muck begun
> Shine, buzz, and fly-blow in the setting sun."

From such an origin, of course, only the vile can emanate and such as have a natural sympathy with corruption. So in metaphor the fly represents, more frequently than any other class, the base-born and base, the maggot-bred and maggot-breeding.

> " Of pension'd patriots and privileged slaves,
>   That party-coloured mass, which nought can warm
>   But quick corruption's heat, whose ready swarm
>   Spread their light wings in Bribery's golden sky,
>   Buzz for a period, lay their eggs, and die."—*Moore.*

> " Those gilded flies that, basking in the sunshine of a Court,
>   Fatten on its corruption ! what are they?
>   The drones of the community ; they feed
>   On the mechanic's labour."—*Shelley.*

> " Oh ! that a verse had power and could command
>   Far far away these flesh-flies of the land ;
>   Who fatten without mercy on the fair,
>   And suck, and leave a craving maggot there."—*Cowper.*

> " Ye tinsel insects whom a court maintains,
>   That count your beauties only by your stains,
>   Spin all your cobwebs."—*Pope.*

Pope, it is seen, mixes up butterfly, blue-bottle, and spider ; but then it is only Pope, and his meaning is obvious enough.

How curious, by the way, that want of generosity is in poets which makes them take every opportunity that offers to remind winged insects of their previous and more lowly stage of existence. One would almost have expected them, as moralists, to see the encouraging aspect of the transformation of caterpillars into butterflies, and grubs into flies, to have taught humility the lesson of hope, and industrious obscurity the gospel of rewards on this side the grave.

The intermediate chrysalis stage is, again, most simply eloquent of patient expectancy based upon a perfect faith. The philosophers of old made much of these beautiful and profound parables, and the many inspiriting readings they offer to the student. But our poets, so rarely as almost to justify my saying, never take heed of them. " The crawling worm that turns to summer fly " is a vermin, a reptile ; and the summer fly only a muck-worm in disguise, braving it like a harlot, in finery for half the day :—

> " Diseased, decayed, to take up half a crown
>   Must mortgage her long scarf and matua gown,
>   Poor creature who, unheard of, as a fly
>   In some dark hole must lie a whole half-year
>   That, for one month, she tawdry may appear."—*Rochester.*

A masquerading maggot, a tawdry gaudy grub. And, say the poets, "blood will out." The fine fly reveals in its progeny its own descent. It is bound to betray itself, and lay eggs which will turn to maggots. All this is very curious, and somewhat puzzling.

However, to return to metaphors. Flies are "courtiers," "sycophants," "beggars," "triflers," "rhymesters," "critics" —everything in fact that poets specially censure, and that are elsewhere described as " vermin " or "reptiles ":—

> " You like the gaudy fly your wings display,
>   And sip the sweets, and bask in your great patron's day."—*Dryden.*

> " Beggars like flys that oft return."—*Broome.*

> " Whether he measure earth, compute the sea,
>   Weigh sunbeams, carve a fly, or spit a flea,
>   The solemn trifler with his boasted skill
>   Toils much, and is a solemn trifler still."—*Cowper.*

> " Witlings, brisk fools, cursed with half-sense
>   That stimulates their impotence,
>   Who buzz in rhyme, and, like blind flies,
>   Err with wings for want of eyes."—*Green.*

The fly-critic deserves more than one quotation, so here are three, from Butler, Young, and Byron :—

> " Critics are like a kind of flies that breed
>   In wild fig-trees ; and, when they're grown up, feed
>   Upon the raw fruit of the nobler kind,
>   And, by their nibbling on the outward rind,
>   Open the pores ; and make way for the sun
>   To ripen it sooner than he would have done.

> Slight peevish insects round a genius rise
>   As a bright day awakes the world of flies ;
>   With hearty malice, but with feeble wing,
>   (To show they live) they flutter and they sting ;

> But as by depredations wasps proclaim
> The fairest fruit, so these the fairest fame.

> Humming, like flies, around the newest blaze
> The bluest of bluebottles you e'er saw,
> Teasing with blame, excruciating with praise."

Youth hovering round dangerous temptation, and desire for beauty, have of course their poetical counterfeits in insects, that flutter into flames and rifle flowers. In the first catastrophe the moth is properly the usual victim, but sometimes the fly, as Ben Jonson's simpleton,

> "In his mistress' flame, playing like a fly,
> Was turned to cinders by her eye."

Or Byron's "youthful friend,"

> " E'en now thou'rt nightly seen to add
> One insect to the fluttering crowd ;
> And still thy trifling heart is glad
> To join the vain and court the proud.
>
> There dost thou glide from fair to fair,
> Still simpering on with eager haste ;
> As flies along the gay parterre
> That taint the flowers they scarcely taste."

A more original fancy than the majority is, however, Mackay's :—

> "And there were other suitors, human flies,
> That ever drone and buzz at honey pots ;
> With busy wings, lank legs, and suckers dry,
> For want of golden sweets ; that long to light
> Upon the paths of widows richly dower'd,
> And settle there ; insatiate as wasps
> That dig their feelers into luscious pears
> Or burrow into peach or apricot."

Mallett, too, finds in the bluebottle the suggestion for an excellent stanza :—

> " Still hov'ring round the fair at sixty-four,
> Unfit to love, unable to give o'er ;

A flesh-fly, that just flutters on the wing,
Awake to buzz, but not alive to sting;
Brisk where he cannot, backward where he can,
The teasing ghost of the departed man."

Is "insect" a good synonym for man? The poets think it is, but I confess I do not agree with them. Not because the word is itself absurd as applied to creatures with solidly continuous bodies like men and women (for poetry does not of necessity concern itself with the real meanings of words), but because I am no pessimist, and think much too well of my kind to allow that *human* beings are either reptiles, vermin, or insects.

But reptiles and insects are words which, outside of poetry and hyperbole, have exact meanings, and there is nothing in the nature of these creatures to make their name a term of reproach, or befitting as a synonym for worthless men.

What has an underhanded, sneaking coward in common with the ineffable splendour of forest-ruling pythons? Where does the comparison begin between the cringing, malice-mongering woman-wretch and the awfulness of the elegant and courageous vipers? Why are the mischief-making and the worthless called by the name of harmless, useful toads? Does the crocodile properly illustrate small, pitiful hypocrisy? No: I cannot admit that the baseness of humanity finds any reflection in reptiles; nor yet in insects. It is, of course, a gross degradation to a human being that he should so lose all appearance of reason in his conduct as to seem no better than an unreasoning insect. But meanwhile it should not be forgotten that in this declension of humanity from its own noble level there is no ground for making the insect sink with the man. Because a filthy man descends to the same pleasure in filth that a dung-beetle finds in the dunghill, is the latter dirty? Because he leaves his trail on everything he touches, is the snail to be abused for defiling? Because he taints whatever he settles on, is the fly as noisome as he? Surely not. The dung-beetle, the snail, and the fly are on their proper plane, and are cleanly, industrious, and respectable.   When a human being delights to burrow and wallow in dirt, to beslime all he touches, to deface and besmear the beautiful, he dishonours his own reason, and becomes like something else which is itself admirable, but for a man to resemble shameful.   It is in the act of resembling, and not the thing resembled, that the shamefulness lies, and where reproach should fall.

Monkeys are as good as men.   And yet, when men behave like monkeys, they deserve to be whipped out of human society.   Does this sound irreconcilable?   To tell a man that he is no better than an ape does not necessarily imply that the ape is worse than he.   It does not convey the compliment that he is as nimble, as clever, as moral, as good a husband and father as the ape.   It simply means that his behaviour does not seem to be controlled by human reason. The ape gets none the worse from the man doing so.   It remains what it is, a very admirable wild animal.

1.

"Little fly,
  Thy summer's play
  My thoughtless home
  Has brush'd away.

2.

Am not I
A fly like thee?
Or art not thou
A man like me?

3.

For I dance
And drink and sing,
Till some blind hand
Brush my wing.

4.

If thought is life
And strength and health,
And the want
Of thought is death,

5.

Then am I
A happy fly,
If I live
Or if I die."—*Blake.*

Or, to call a man "a regular toad" is really to say that he is a remarkably sagacious, patient, hard-working, and very useful creature. But what is meant by the reproach is, that he is of a bloated, squat, and objectionable appearance, unsociable, holekeeping. Well, even I do not think (in spite of what those who have kept toads as pets have to say) these creatures delectable in person—and I know persons who look very like toads, just as one knows people with profiles like sheep, horses, baboons, and so forth. But I certainly should not flatter any toad-resembling person if I found him stupid, or irritable, lazy, or mischievous, by saying he was a " regular toad."

However, all this by way of digression, a "spontaneous combustion;" and to return to the original fact, which is, that poets, as a class, are of the opposite way of thinking ; and that when they detect a resemblance between a human act and an insect one, they transfer to the latter all the rest of the former's weaknesses, failings, and vices.

In direct association with man, the most familiar one is the fly that will not let you sleep. "I fain had slept but flies would buzz around" is not very poetical, but it is a very truthful incident. Another is that which gets into the wine you are drinking, and finds a Duke-of-Clarence death therein. Whence Quarles has this moral—

> " The sun-delighting fly repairs at first
> To the full cup ; only to quench her thirst.
> But oftentimes she sports about the brink,
> And sips so long till she be drowned in drink.
> When wanton leisure shall present thine eye
> With lavish cups : Remember but the fly."

And then there is the bluebottle that bangs into your room on sultry afternoons and bangs out again—a theme to which the faithful Hurdis rises :—

> " At the door
> Enters the flesh-fly, and with cheerful hum
> Travels the house interior : on the pane
> Thumps he and buzzes ; the resounding hall
> Travels again ; and with a bounce departs:

> Grateful remembrance leaving on the mind
> Of still enjoyment in the musing hour
> Of summer's drowsy noon, and pleasing thought
> Oft interrupted by his brisk career."

And that other fly that settles on your face, and when you slap it does not get slapped :—

> " Impatient at the foul disgrace
>   From insect of so mean a race ;
>   And, plotting vengeance on his foe,
>   With double fist he aims a blow :
>   The nimble fly escaped by flight ;
>   And skipp'd from this unequal fight.
>   The impending blow with all its weight
>   Fell on his own belovèd pate."—*Somerville.*

The urchin on his way to school stalks the contemplative bluebottle on the sunny wall with all the precautions a Red Indian would spend over a bear ; and, even then, the sudden whisk with which his hand goes all ascrape along the wall, proves as often as not unavailing.   Either the fly was washing his face, and did not get up off the bricks soon enough, or else he was watching the urchin, as the wily bear often watches the Redskin, and was off too soon.   But, whatever the reason, the fact remains that the bluebottle does not.

Moreover, in flying away, it startles all the other flies for yards around, and the happy hunting grounds which the shiny-faced youth had found are suddenly desolated by his single misadventure, just as a prairie is emptied of its bison by the mischance of a rifle going off by accident.

When grown up, the same boy continues to catch flies, or tries to.   He buys gummy compositions, which he smears upon paper, with the expectation that the insects will sit upon it and become entangled on the viscous surface and expire there.   But he is doomed to disappointment wholesale, for the flies come and make prodigious meals off the gummy composition, and when they have done they walk round the edge of the plate with toothpicks in their mouths, as proud as old gentlemen in white waiscoats who have just dined at a grand restaurant.

Some of them eat too much, but they are not so mean-spirited as to show it, or to advertise the stuff that kills them ; for they creep away into remote places—tobacco-jars, jam-pots, tea-caddies, the family Bible, anywhere—and turn over flat on their back, and fold all their six legs across their breasts, and die with the most decent composure possible.

There is no triumph to be scored off such decease. The small things confided in man, and got poisoned for doing so. They thought, as it was intended they should, that the sticky papers were laid out for their refreshment, and partook of the cold collation with that artless trustfulness which, after all, is one of the most agreeable features of the wild world. But the adhesive stuff, neither liquid nor solid—an amphibious sort of mixture, so to speak—had been laid on guilefully, "and the same with intent to deceive," and the flies therefore were as deliberately betrayed to their death as those guests of Jung Bahadur's who were invited to supper, but beheaded before they had time to begin. As a specimen of hospitality, the treacled plate is execrable ; as a practical joke, it is in the worst possible taste.

Thus far, then, the association of flies with human beings which in the poets is so characteristically prominent but in nature rather accidental, casual, and secondary. Far more important in nature is the association between the fly and spider. Each is the complement of the other ; each the corollary to the other's proposition. From spider you deduce fly ; it is arguing from the known to the known. The one is as implicit in the other as bride in bridegroom. You cannot imagine flies without spiders any more than you can imagine matter without form. They go together like sub-stance and shadow. The idea of either alone is an absurdity, a negative quantity. Subtract flies from spiders and what remains ?   And this intimate relativity is not lost sight of by the poets.   The literature of fly-spiderism is immense.

Their Fate sits toiling for the flies everywhere, three Fates in one, spinning, measuring, and cutting into lengths ; and there is a mathematical neutrality, a cold, calculated deliberateness about the spider as it works that is significant of assured success.   How it tries each thread as it is finished : how punctually it halts at each crossing of the strands to secure the line : with what an air of "Now I'm ready for dinner," it takes its seat.

Yet there are many delightful passages in poetry recognising the fact that occasionally the silly fly does *not* walk into the crafty spider's parlour to gossip, and that it *does* sometimes break through "the flimsie nette," and make a triumphant escape, leaving the spider with a great deal of temper to make good, and a large hole in the net as well. Thus Bloomfield enlarges on a bluebottle's scrapes.   It had escaped, it says, from a sparrow and then from a man's hand, but flew "with such adour and glee" that it went headlong into a cobweb, with its owner at home :—

> " Who so fiercely came out
>   Of his hole, that no doubt
> He expected that I was secure.
>   But he found 'twould not do ;
>   For I found my way through,
> Overjoyed at escaping, you're sure."

Another association, hardly more agreeable to the fly, is that with fish.   As a rule they perish of their own act, when, " on the sunny shallows resting," they "tempt the watching fish to spring," or when, as in Grahame, they provoke two Fates in one :—

> " Dimpling the water glides, with here and there
>   A glossy fly skimming in circlets gay
>   The treacherous surface ; while the quick-eyed trout
>   Watches his time to spring ; or, from above,

> Some feather'd dam, purveying 'mid the boughs,
> Darts from her perch, and to her plumeless young
> Bears off the prize."

That birds should eat flies does not, somehow, affect the poets. The spider is a crafty villain and a murderer, but "the sweet songster" that carries off insects to its "callow young" is only fulfilling a beautiful parental function. This is, of course, in a way as it should be, for if it were not for perjudices poetry would often be but flat stuff; yet it is worth noting in passing. Owls are formally damned as fiendish for singing to their mates at night; not so nightingales. Storks are complimented on eating frogs, but a wolf must not even look through the fence at mutton.

Hitherto the fly has figured only in its sadder aspects, as the creature of corruption and an emblem of the corrupt, the tormentor of man, the victim of spiders, the prey of fishes and of birds. But it has its gayer side as the persecutor of cattle and sheep, as what Hiawatha calls

> " The stinging fly, the Almo,"

but which our poets address as "the breeze." The poets positively revel in this insect. For herds and flocks are especially their delight; and as it is impossible in summer to contemplate either one or the other without becoming aware of "the breeze" at work among them, the fly is worth whole volumes.

Both sheep and cattle are of a monotonous sort to watch. Unless outside influences operate, they are a dull lot. It is not often they originate any excitement. Their *summum bonum* is a complete placidity; their idea of Elysium broad meadows where some bird of the halcyon kind should brood perpetually. But even on poets such unruffled scenes would pall in time. Even Cowper and Wordsworth were annoyed by the uneventful tranquillity of cow-happiness. So "the breeze" comes as a boon and a blessing.

The vagrant cur is acceptable as an incident; the stranger animal from another herd is welcome; the calf a god-send to "milky-mother" verse. But "the breeze"! What animation it at once imparts to the scene. "Teasing fly-time"! At ordinary times the herds are going to pasture or coming back, eating or reposing. They chew the cud and get milked. The heifer's bell tinkles drowsily at times. A cow in the lane says "Moo." But the life is dull, say what you will; and if it had not been for "the fly that pricks the gadding neate," I cannot imagine what some of our poets would have done for pastoral verse. No set of cows, let them be ever so variegated in colour, could well be sung of more than once if they only grazed and lay down. As a touch of spring in a rural poem, the "placid beeves," "unworried in the meads," are admirable once in a way and for all; and "the calm pleasures of the pasturing herds" complete the vernal scene very handsomely—if not recurring too frequently.

> "The cattle are grazing,
> Their heads never raising;
> There are forty feeding like one."

But when summer comes "the breeze-fly" comes with it, and great is the relief of poets. "Good-bye to the balm, of palpable and breathing calm." The herds are now "restless." They rush from their tormentors into the water, and "standing knee-deep, scare off with sudden head reversed the insect swarm," or "fly with tufted tail erect to the shade," "scampering madly." The whole field is astir with swinging tails. Never still, the herd offers a perpetual succession of incidents to the poets, a kaleidoscope of cattle. Feeding becomes a running fight with gad-flies, and even he tranquil milking episode a skirmish

> "Even at the close of day
> Unruly cows with marked impatience stay,
> And vainly striving to escape their foes,
> The pail kick o'er."—*Bloomfield.*

Nor do the sheep fare much better, "what time the new-shorn flock stand here and there, With huddled head, impatient of the fly." For the shearers are just then at work, and

> " The fretful ewe that moans an equal share,
> Tormented with flies, her head she hides,
> Or angry brushes from her new-shorn sides,"

is a welcome touch of summer, which the poets gratefully acknowledge by frequent use.

> " Fell Œstrus buries in her rapid course
> Her countless broods in stag, or bull, or horse ;
> Whose hungry larva eats its living way,
> Hatch'd by the warmth, and issues into day."—*Darwin.*

The horse, too, with cruelly cropped tail, commands the sympathy that is due.

> " By the unclouded sun are hourly bred
> The bold assailants that surround thy head,
> Poor patient Ball ! and, with insulting wing,
> Roar in thine ears, and dart the piercing sting.
>
> In thy behalf the crest-waved boughs avail
> More than thy short-clipt remnant of a tail,
> A moving mockery, a useless name,
> A living proof of cruelty and shame.
> Shame to the man, whatever name he bore,
> Who took from thee what man can ne'er restore ;
> Thy weapon of defence, thy chiefest good,
> When swarming flies contending suck thy blood."

Why are flies so unpopular? That everybody dislikes them everybody knows. Luther hated them, and massacred them without mercy. He said they were "emissaries of Diabolus, and the ghosts of heretics," because whenever he was reading a pious book they paraded about upon it to distract his attention, and soiled it. Long before Luther's time, however, they were specially affiliated upon Beelzebub, the patriarch prince of bluebottles. The monks abominated them, and said they were immoral. Religious legends of the Talmud are to the discredit of the dipterous vagabond.

The Mussulman brings his slipper down on a fly "in the name of the Prophet." In hot countries special engines are prepared for their discomfiture and destruction—prodigious whisks of horsehair or yaktail, round flaps of leather attached to long handles of cane. Sancho Panza cursed them as being enemies to sleep; and all through Southern Europe they are under the ban of a universal execration. "Fly-time" is in half the world a season of terrors; when commerce hesitates to busy itself, social arrangements are in abeyance, and everything is dislocated and in disorder, simply because the flies are abroad.

One of the plagues of Egypt was the fly. It is one of the penalties of Purgatory. All this is, of course, very much to the discredit of this small satellite of man, this importunate dependent of humanity.

Historically, flies are insignificant. In Philistia they had a fly-god, Baalzeebub. Egypt, in her ancient litanies, prayed to Achor for deliverance from them, but, judging from modern Egypt, with but scant response. Cowley, in his ode, makes the mistake of thinking Aaron's plague was a miscellaneous assortment of species, a mixed entomology let loose wholesale upon the Pharaoh and his people.

> " Harmful flies, in nations numberless,
>   Composed the mighty army's spurious host ;
>   Of different manners, different languages,
>   And different habits, too they were,
>   And different arms they bore ;
>   And some, like Scythians, lived on blood ;
>   And some on green, and some on flowery food.
>   And Acharon the airy prince led on this various host."

Now, Cowley, thinking to improve on the original, has destroyed the whole horror of the plague ; for surely there is something positively grotesque in a various host of wasps, gad-flies, hornets, dragon-flies, bluebottles, bumble-bees, fire-flies, mosquitoes, may-flies, gnats, sand-flies, midges, and all the rest of them.

The real, overwhelming, loathsome horror of the visitation was of course this, that the land of Egypt suddenly swarmed from end to end with *house-flies*, and no others.   They did not sting, nor bite.   They did nothing aggressive, but simply sat in sheets, in heaps, everywhere, acres of them, square miles, crawling one over the other, ever-shifting clouds, almost too thick to walk through, perpetually rising and re-settling.

Who that has been in Egypt in the hot weather has not felt the fly an almost intolerable burden, a presence almost too nauseating for endurance ?   And the bazaars !   Even un-plagued they are a memory to shudder at.   What is that man yonder selling ?   As some one passes, the black plaster of flies lifts heavily for an instant off the wares on his stall. They are ruddy in colour.   What are they ?   Sweetmeats dyed with pomegranate juice ?   Water-melons split to show their rosy freshness ?   Or meat ?   The seller is asleep in the corner, his clout over his head, and the flies hang in bunches from every stain on the dirty rag.   And worse, and worse, and worse is seen, till the cumulative horror would shame an English page to describe it, and sicken the reader.   So why did Cowley—he often does it in the same poem—think to improve upon Divine vengeance, so simple and yet so inconceivably shocking, by the elegancies of variety ?

Moore has a poem on the Egyptians' worship of the hosts of Achor, in which he turns the point to suit his own political prejudices :—

> " The wise men of Egypt were secret as dummies,
>   And even when they most condescended to teach,
>   They packed up their meaning as they did their mummies
>   In so many wrappers 'twas out of one's reach.
>   They were also good people, much given to kings,
>   Fond of monarchs and crocodiles, monkeys and mystery,
>   Bats, hieophants, bluebottle flies, and such things :
>   As will partly appear in this very short history.

> A Scythian philosopher (nephew they say
> To that other great traveller, young Anacharsis)
> Stepped into a temple of Memphis one day,
> To have a short peep at their mystical farces.
> He saw a brisk bluebottle *fly* on an altar ;
> Made much of and worshipped as something divine ;
> While a large handsome *bullock*, led there in an halter,
> Before it lay stabbed at the foot of the shrine.
>
> Surprised at such doings, he whispered his teacher,
> If 'tisn't impertinent, may I ask why
> Should a bullock, that useful and powerful creature,
> Be thus offered up to a bluebottle fly?
> No wonder, said t'other, you stare at the sight,
> But we, as a symbol of monarchy view it :
> That fly on the shrine is Legitimate Right,
> And that bullock, the people that's sacrificed to it."

The fly that sat on the wheel and prided itself on the dust that was raised, the other that flew up with the eagle and nestled in its eyrie, and Io's bane, complete, I think, the poets' record of legendary flies of note.

Yet, who admires the fly? It is true that Homer compares the valiant Greeks to a fly ; and never was simile more apt. For what can exceed the astonishing courage of this insect, the reckless intrepidity of its assault, or the desperate persistence of it? Supposing, as some one says, a man were out walking, and a seven-acre field suddenly turned upside down with him. For this is exactly what happens to a fly every time you whisk it off with your hand. But it comes back exactly to the same spot ! What man of us would do as much? It is true the fly has made itself familiar with such sudden upheavals of an apparently solid surface, and this argues no trifling degree of nerve and resource. If the thing were a blockhead and a dunce, and got killed for its clumsiness every time it sat down, it would be another affair altogether ; and the bluebottle would be only a kind of Mr. Feeble who gave in to the first giant he met. But this is not so, for in the matter of lives it takes about nine cats to make one fly. The insect graduates in adventure like the

Student of Santillane, accepts the most appalling disasters of existence with the indifference of Sinbad, and treats bodily peril with the lofty scorn of Don Quixote.

The fly in fact is an expert in the evasion of sudden death. It is assailed by the equivalents of thunder and lightning, of cannon-fire and volcanic explosion, but escapes them all. Dynamite is sprung upon it without avail. It laughs to scorn the shaking of the spear. Honest hostility in fact is of no use. It would not care in the least for all the king's horses and all the king's men. But against treachery what courage is of avail? Beset by the blandishments of a false friendship, what heroism can be proof? So the fly finds its end multitudinously in poisoned treacle, and the insect that would have braved, if necessary, the thunders of Assaye, falls a victim to the sticky insidiousness of the catch-em-alive-ohs. Whether this is as it should be is for the judgment of each individual.

But besides " the fly " ordinary, there are other species which the poets mix up with the domestic insects. Thus the "bluebottle" becomes the gad-fly when we find it "having tormented man, urge unsatisfied its course to torment the beast." As a matter of fact

> " All the race of silver-wingèd flies
> Which do possess the empire of the air "

are lumped together—which perhaps is no more than is justifiable, for after all, if poets became entomological, verse would suffer. Still it is not justice, not even courtesy, to nature.

The gad-fly, however, has a very distinct individuality. It is "fell Œstrus ; " "the maddening fly," "the humming gad-fly," which (like the critic and author) "imprints its malicious comments on the tender flank." In summer it is of course supreme in bovicultural verse :—

> " And scorching sunbeams warm and sultry creep
> Waking the teazing insects from their sleep :
> And dreaded gad-flies with their drowsy hum
> On the burnt wings of mid-day zephyrs come,
> Urging each clown to leave his sports in fear,
> To stop his starting cows that dread the fly :
> Droning unwelcome tidings in his ear
> That the sweet peace of rural morn's gone by."—*Clare.*

The May-fly, always pitied as being ephemeral and the prey for fishes, is often, and very charmingly, noticed as "fluttering, for a summer's day, upon the glassy bosom of the pool," or (Savage) "dancing on the stream till the watery racer snatches it away." It is a regularly recurrent feature of spring evenings as the "quick water-fly" or "mazy insect." But, nevertheless, well as the poets knew it, it is constantly found maggot-bred, living in courts or on popular favour, and so forth, "for a day."

The dragon-fly is, curiously enough, a great favourite with the poets. Every reference to it is admirable, as, for instance, these examples :—

> " And forth on floating gauze, no jewell'd queen
> So rich, the green-eyed dragon-flies would break
> And hover on the flowers, aërial things,
> With little rainbows flickering on their wings."—*Jean Ingelow.*

> " And just above the surface of the floods,
> Where water-lilies mount their snowy heads,
> On whose broad swimming leaves of glossy green
> The shining dragon-fly is often seen."—*Clare.*

> " Mark how clear
> It sparkles o'er the shallows ; and behold
> Where o'er its surface wheels with restless speed
> Yon glossy insect ; on the sand below
> How the swift shadow flits."—*Southey.*

Of course it is obvious that, except the naturalist Darwin, who sees

> " Fierce Libellula, with jaws of steel,
> Ingulf an insect province at a meal,"

none of the poets really understood what the dragon-fly, the dispeopler of the air, the tyrant of the pool, really is.   They do not recognise in the creature that

> " with gauzy wings,
> In gilded coat of purple, green, or brown,
> That on broad leaves of hazel basking clings
> Fond of the sunny day,"—*Clare.*

the vulture, the shark, the wolf—the everything that is poetically dreadful to the insect world—that it is a carnivorous fly and singularly pitiless.   Perhaps it is better that they did not know it, for the insect, as it is, has adorned their verse, and not suffered in the handling.

The " winged ichneumon " for her embryon young, that " gores with sharp horn the caterpillar throng," is also the " false ichneumon."   The gnat is in Hurdis "the minor fly " :—

> " If yet the season to his race be kind,
> Sharp stings the minor fly, chirurgeon keen,
> With lancet petulant, the manly skin
> Provoking, oft repuls'd, and slaking well
> His thirst of blood, ere the vindictive hand
> Of his vex'd patient fall, and with a frisk
> The small phlebotomist indignant crush."

As diverse as the winged insects are the " maggots " and " grubs " that breed them.   But in their perfected and imperfect forms they are employed equally indiscriminately as typical of degraded humanity.   Thus the meal-worm, " all powdered o'er from tail to chin " (in Swift), " cheese-hoppers," bookworms (as in Burns) :—

> " Through and through the inspired leaves,
> Ye maggots, make your windings ;
> But oh ! respect his lordship's taste,
> And spare his golden bindings."

Apple-grubs that in the cider-press

> " Oft unobserved invade the vital core,
>   Pernicious tenants ! and their secret caves
>   Enlarging hourly prey on the pulp."

The "gentle" (as in Clare) :—

> " For make-shifts oft crook'd pins to thread were tied,
>   And delve his knife with wishes ever warm
>   In rotten dunghills, for the grub and worm
>   The harmless treachery of his hooks to bait."

And many another, too,

> " Not seen but understood,
>   That live in vinegar and wood."

Only once, however, is the "maggot" defended, and that is in Southey's "Filbert" :—

> " Nay, gather not that filbert, Nicholas ;
>   There is a maggot there, it is his house,
>   His castle, oh ! commit not burglary ;
>   Strip him not naked ; 'tis his clothes, his shell,
>   His bones, the very armour of his life ;
>   And thou shalt do no murder, Nicholas."

To be "enkernelled" thus, regardless of the daily papers and all the turmoil of life, must be delightful :—

> " The perfection this
>   Of snugness ! It were to unite at once
>   Hermit retirement, aldermanic bliss,
>   And Stoic independence of mankind."

CHAPTER V.

GRASSHOPPERS, CRICKETS, AND LOCUSTS.

LIKE the nightingale or the frog, the grasshopper lives in verse chiefly from its "song." It is the minstrel among the insects; the "piping" one. Not that all admire it, for it is called "tiresome," "shrill," "creaking;" while Marvel speaks of its note as a "squeaking laugh." Nor is this, on occasion, a bad description; for I remember once, when puzzled for my way on the Wiltshire Downs, fancying the cricket's voice derisive. I was listening with both my ears to catch if I could the sounds from some neighbouring sheep-fold; but all I heard was the jeering of the insect. But, like church-bells, grasshoppers say just what the listener chooses to hear. So its speech would work in well into fairy-tale as a substitute for the ambiguous directions given to straying heroes and heroines by mischievous elves, or that old man's nod of the head, which in Red Indian stories means yes or no, just as the inquirer wishes the answer to be the one or the other.

"Twittering" is more than once the poet's epithet for the sound (Leyden has "pittering," which, I suppose, is a phonetic rendering), and, as a rule, it is amiably accepted as after the manner of "singing." Mackay is original—

> " By the clink that sounds among the grass,
> Like tempered steel on greaves of brass,
> As the mail-clad grass-hoppers chirp and pass."

Byron has it "carolling," and others call it a "blithe singer"
cheering the mower and the rustic at their tasks.   Both
Leyden and Lovelace are worth quoting, but what the latter
means is past comprehension.

> " And pittering grasshoppers, confusedly shrill,
>   Pipe giddily along the glowing hill ;
>
> Sweet grasshopper, who lov'st at noon to lie
> Serenely in the green-ribbed clover's eye,
> To sun thy filmy wings and emerald vest,
> Unseen thy form, and undisturbed thy rest !
> Oft have I, listening, mused the sultry day,
> And wondered what thy chirping song might say ;
> When nought was heard along the blossomed lea
> To join thy music, save the listless bee."

> " Oh ! thou that swing'st upon the waving hair
>   Of some well-filled oaten beard,
>   Drunk ev'ry night with a delicious tear
>   Dropp'd thee from heaven.
>
> The joys of earth and air are thine entire,
> That with thy wings and feet dost hop and fly,
> And when thy poppy works thou dost retire
> To thy carved acorn bed to lie."

"Hoarse" is an epithet frequently applied, but, as a rule
—for instance Southey's,

> " Hoarse grasshoppers their evening song
>   Sang careless as the dews of night descended,"

and Leydon's,

> " The tuneful birds suppress the cheerful lay,
>   And to hoarse grasshoppers resign the day,"

in relation to the insect at nightfall.   Is there really any
change in its tone, as in that of frogs ?   I have myself
thought that its vespers differed from its matins in being
deeper and less sprightly.   At any rate, it is certain that the
creature constantly modulates its tones, and, listening to it,
it seems as if the small scraper got tired or inattentive ; its

"voice" becomes irresolute, flat, false in quantity; then it pulls itself together and starts afresh clear, firm, and true. It is very easy, if you have the whim to do so, to guess at some of the grasshopper's moods from the way it sings.

It is worth, perhaps, saying that the grasshopper makes its music by scraping the inner edge of its long hind legs against the ridges of the wings, and the cricket by rubbing the wing-covers against the wings.

> " The cricket chirrup'd in his coat of mail;
> The brisk cicada answered him aloud
> And rubb'd the emerald armour of his thighs "

Why they should make these noises is not obvious; but as it is only the male that is harmonious, the presumption is that the music serves the same purpose as analogous accomplishments in birds—as being an ornament to the "stronger" sex and an attraction to the "weaker." The poets, by the way, almost invariably address the chirping grasshopper as " she," just as they always transfer the male nightingale's song to the hen-bird.

Among the oddities of zoological folklore I find the grasshopper written down as an idle and thoughtless person. It is of a loquacious kind, a chatterer, and therefore flighty, irresponsible, a ne'er-do-weel. It starves when hard times come; begs its bread in winter. So it figures as the opposite of the ant; a contrast to the silent and industrious emmet. How old the idea may be no one can say; but, at any rate, it is as venerable as the most ancient Sanskrit legends. For there we find the grasshopper spoken of as an improvident individual and an unreliable. It runs a race with the ant, but after making some astonishing leaps takes a nap, just as the hare does when racing the tortoise, and of course the ant plods in first. Again, it neglects to store its larder, and the ant—a detestable little prig in folklore—gives it a good lecturing when it ought instead to have helped the poor

mendicant with a good grain of wheat.   Æsop reproduced
the idea, and the first of La Fontaine's fables is a version of
the Phrygian's

> " A grasshopper gay
> Sang the summer away,"

and, having nothing to eat in winter,

> " A begging she went
> To her neighbour the ant,"

who asks her what did she all the rest of the year that she
now finds herself in such wretched case.   The grasshopper
confesses she sang for all comers.   "Sang indeed!" is the
emmet's unsympathetic reply, " Then now you must dance."

> " The remiss
> Dimetians, who along their mossy dales
> Consume, like grasshoppers, the summer hour,
> While round them stubborn thorns and furze increase,
> And creeping briars."—*Dyer.*

In the poets the same fable frequently recurs.   Spenser
starts it in the "Shepherd's Calendar"—

> " And my poor muse hath spent her spare store,
> Yet little good hath got, and much less gain,
> Such pleasaunce makes the grasshopper so poor,
> And beg for bread when winter doth her strain,"

and others repeat it both of the grasshopper and the cricket.

Now this fancy is more curious than at first appears; for
this reason.   There is actually a grasshopper *which lives
with ants*, and shares their home with them, under some
such queer arrangement for reciprocity as makes it possible
for owls and snakes to share the marmot's dwelling.   Upon
what terms the ant tolerates the grasshopper does not appear,
or how the grasshopper justifies its partnership.   Yet there
it is.   The *modus vivendi* has been found; and incongruous
as the association seems, it must evidently have some
recommendations that satisfy both parties, or else it could
not exist.

And failing fact, why not let fancy suggest an explanation ? The ants find life dull, the ceaseless round of silent and prosaic drudgery intolerable.   So they pay with board and lodging the minnesingers of the meadows, the troubadours of the stubble, to live with them and cheer the toilers of their austere republic with music.   Thus too do we see men of monotonous labour sitting at their work with singing-birds hanging overhead.   In the old sailing-ship days the fiddler fiddled to the crew when at work.   Music always lightens labour of a mechanical kind.   What new life the band puts into the legs of marching soldiers !   In every country in the world you will find the coolies, the porters, and those who perform tedious and unvarying work, invariably chanting as they toil.   Overseers of all kinds encourage it.   The fact is as old as the hills—as ant-hills.   And so too (there is no harm in such imaginings) the industrious emmet may perhaps, after all, not be such a fanatic in the unwisdom of perpetual slaving as has been thought.

But if my fancy be allowable, that the ants (do they not keep wood-lice as housemaids ?) entertain musicians for the brightening of their daily lives, the poet is at fault, and the cricket's scraping "relieves the o'erlaboured bulk " and teaches a "seasonable smile."   All of which would be very engaging—for I consider the ceaseless moiling of ants abominable.

Let this be as it may, the fact remains that grasshopper-folk and ants do live together, and so the bottom falls delightfully out of the fable which makes the one intolerant of the other's idle habits.   The grasshopper, the singer of an idle hour, finds his friends in the appreciative citizens of the little hills, and the pismire is no longer an unsympathetic little prig.

Curious, too, is the notion that grasshoppers are short-lived : is it from these creatures that the phrase "a short life and a merry one " is taken ?

Of course it is pre-eminently a summer insect.

> " Oh, it is pleasant in this summer time
> To sit alone and meditate a rhyme,
> To hear the bee plying his busy trade.
> O ! grasshopper alert in sun and shade,
> With bright large eyes and ample forehead bald,
> Clad in cuirass cuishes emerald."—*Mackay.*

Who does not remember the day in July when, out in the meadow, or among the heather, or on cliffs that overlook the sea, or in some woodland corner in the country, the grasshopper's voice "filled every pause"? How the little creature's chirrup-chirrup reinforces the idea of a sultry midsummer day ! What drowsiness it lends to it !

> " The forest deep,
> That drowsy rustled to the sighing' gale ;
> And still a coil the grasshopper did keep ;
> Yet all these sounds yblent inclinéd all to sleep."—*Thomson.*

In poetry this small lyrist " that lives on noonday dew "—whatever that may be—is a favourite figure in nearly every description of hot midday, whether the poet be one who is faithful to facts from familiarity—like a Bloomfield, Clare, or Grahame—or one who, like Marvel, Keats, or Shelley, are always in fancy so delightfully in sympathy with the spirit, if not always the letter, of nature's doings :—

> " The poetry of earth is never dead ;
> When all the birds are faint with the hot sun,
> And hide in cooling trees, a voice will run
> From hedge to hedge about the new-mown mead :
>
> That is the grasshopper's—he takes the lead
> In summer luxury—he has never done
> With his delights, for when tired out with fun,
> He rests at ease beneath some pleasant weed."—*Keats.*

> " Come, be happy ! lie thee down,
> On the fresh grass newly mown,
> Where the grasshopper doth sing
> Merrily—one joyous thing
> In a world of sorrowing."—*Shelley.*

Marvel too, among his meadows, is just as sympathetic :—

> " Oh, what unusual heats are here
>   Which thus our sun-burn'd meadows fear !
>   The grasshopper its pipe gives o'er,
>   And hamstring'd frogs can dance no more ;
>   But in the brook the green frog wades,
>   And grasshoppers seek out the shades ;
>   Only the snake, that kept within
>   Now glitters in its second skin."—*Marvel.*

Or the more naturalistic poets :—

> " It is high noon,
> And scarce a chirping grasshopper is heard
> Through the dumb mead.   Distressful nature pants."—*Thomson.*

> " Intense the viewless flood of heat descends
>   On hill and vale and wood and tangled brake,
>   And to the chirping grasshopper the broom
>   With crackling pod responds."—*Grahame.*

Granted, then, that it is a summer insect—that it is in the days of heat one hears it most loquacious—how odd it is that the fable should have survived that it " died in October "! It does not, of course, do anything of the kind, but simply withdraws into its place and sleeps through the winter.   As every one knows, it is a rule of nature every winged insect shall die within the year (the occasional individuals that survive the twelvemonth only proving the rule), for the stage of wings is the last third of the creature's life.

After all, it would be very absurd if we did not recognise among ourselves the stages of childhood, youth, middle age, and old age, which together cover the span of our "three score years and ten."   An insect's stages proceed in a far smaller compass, and the winged one is the last.   It is really *the old age* of the caterpillar or grub.   Thus a grasshopper may be for two or three years a grub, for another six months a hobbledehoy—that is, a wingless thing, half grub, half grasshopper—and then for a further space a winged grasshopper.   In the last stage it marries, and there is an end of its purpose.   Nature has no further need for it, and does not care whether it dies or not.

The slender fragility of the insect's appearance may have suggested a feeble hold of life ; some grasshoppers look like the mere spectres of insects.   About others too there is a vegetable, perishable look, as of thin grass-blades that a frost would kill or heat shrivel up; a suspicion about their sere and faded edges that they are already beginning to wither.   But the grasshopper has nothing to complain of as to its length of life.   It sings the summer in and the autumn out, and goes to sleep with the year.

The cricket, the Pau-puk-keena of Hiawatha, "shrill and ceaseless," differs but little from the grasshopper :—

> " Beside yon pool as smooth as glass,
>   Reflecting every cloud,
>   Securely hid among the grass,
>   The crickets chirrup loud."—*Clare.*

And again, in Clare—

> " In mid-wood silence, thus, how sweet to be,
>   Where all the noises that on peace intrude
>   Come from the chittering cricket, bird, and bee,
>   Whose songs have charms to sweeten solitude."

As a matter of fact this insect is, I fancy, only the grass-hopper over again (Keats goes afield to hear the "dappled" cricket), but used under another name for the sake of variety, for it not only "pipes," sings, "chirps," and (in Clare) "chitters and twitters" o'er its dreams, and (in Leyden) even "pitters," exactly like the grasshopper, but has all other points in common with it; and besides there is no very common out-of-doors cricket in England.   Shenstone describes its voice as "tinkling," and Cowper calls it a "locust" in his appeal to the swallow :—

> " Attic maid ! with honey fed,
>   Bearest thou to thy callow brood
>   Yonder locust from the mead,
>   Destined their delicious food ?

> Ye have kindred voices clear,
>   Ye alike unfold the wing,
> Migrate hither, sojourn here,
>   Both attendant on the Spring.
>
> Ah ! for pity drop the prize ;
>   Let it not with truth be said
> That a songster gasps and dies,
>   That a songster may be fed."

Once, and once only, in Garth, I find it called "sinister," which is quite in keeping with homely superstitions. As a rule Clare's domestic cricket that "by the fireside unmolested sings "—

> " Blithe as the lark, as crickets gay,
>   That chirrup on the hearth—"

finds more distinctive notice, and is universally a favourite. "Crickets chirrup on the hearth, As if they shared the children's mirth." The poet of the Pleasures of Melancholy is "blest with the lowly cricket's drowsy dirge ;" but Barry Cornwall, referring to contemporary credulities, calls it the "chilly midnight cricket" with "warning" voice. The idea—and one not altogether without foundation—that the superior comforts of fireside life lengthen the insect's life, is often hinted at, and in the following from Charlotte Smith, explicitly set forth :—

> " Little inmate full of mirth,
> Chirping on my humble hearth,
> Wheresoe'er be thy abode,
> Always harbinger of good.
>
> Though in voice and shape may be,
> Form'd as if akin to thee,
> Thou surpassest, happier far,
> Happiest grasshoppers that are ;
>
> Their's is but a Summer song,
> Thine endures the Winter long,
> Unimpair'd, and shrill, and clear,
> Melody throughout the year.
>
> Neither night nor dawn of day
> Puts a period to thy lay."

Underlying this, of course, is that pathetic main idea of the poets that "life's a short summer." Men and women are mere insects, "the summer swarm." So the poets, whenever they meet with a beauty—the song of birds, the butterfly's colours, the glowworm's spark, a flower in its prime—see in it that which is transient, futile, doomed. Both gaiety and merriment have in verse melancholy significances; night, winter, death, hold them in reversion. Let the grasshopper chirp: it will die soon.

> " Or, if a different image be recalled
>   By the warm sunshine, and the jocund voice
>   Of insects chirping out their careless lives,
>   On these soft beds of thyme-besprinkled turf,
>   Choose, with the gay Athenian, a conceit ;
>   A sound blithe race ! whose mantles were bedecked
>   With golden grasshoppers,[1] in sign that they
>   Had sprung, like those bright creatures, from the soil."
>   — *Wordsworth.*

A very different creature is the locust, "the scourge of Allah," "the army of the One God."

" A fire devoureth before them ; and behind them a flame burneth : the land is as the garden of Eden before them, and behind them a desolate wilderness ; yea, and nothing shall escape them.

" Like the noise of chariots on the tops of mountains shall they leap, like the noise of a flame of fire that devoureth the stubble, as a strong people set in battle array."

Just as Job exhausted for all time to come the poetry of the impregnable majesty of an individual strength in his picture of Leviathan, so Joel, in the chapter of which I have quoted two verses, exhausts the poetry of the irresistible might of multitudes. No poet has ever bettered by a single

---

1 " Witness that Royal Bourse he bade arise,
    The Mart of merchants from the East and West ;
    Whose slender summit, pointing to the skies,
    Still bears, in token of his grateful breast,
    The golden grasshopper, his chosen crest."—*Hood.*

thought the verses of Job, nor any after him added a force or a beauty to the lines of Joel.

The locust has but one aspect in poetry—that of a multitudinous evil—

> " As, borne by winds along, in baleful cloud,
> Embody'd locusts from the wing descend,
> On herb, fruit, flow'r, and kill the rip'ning year,
> While waste behind, destruction in their track,
> And ghastly famine wait."—*Mallet.*

They serve, therefore, as a simile for anything that desolates or devours—"Gaul's locust host," or any other enemy of Britain or of "Freedom ;" armies of all kinds ; the minions of tyranny ; corrupt courtiers ; Jesuits. They are "tree-blasting," "sky-clouding," "blackening all the ground," "in darksome clouds," "hosts that desolate the earth and dim the day," "barb'rous millions," "greedy troops," "endless legions on sounding wings," "thick-phalanxed as when plaguing Samarcand," "dire with horrid swarms." Nearly every poet at one time or another has told "what deeds of woe the locust can perform ;" but their language toils in vain after the consuming, overwhelming reality such as the prophet saw it from the mountain side in Palestine—"a day of clouds and thick darkness—a great people and a strong."

Moore is ribald, of course, on a theme so serious :—

> " Then erst could Egypt, when so rich
> In various plagues, determine which
> She thought most pestilent and vile
> Her frogs, like Benbow and Carlile ?
> Croaking their native mud-notes loud
> Or her fat locusts, like a cloud
> Of pluralists, obesely lowering
> At once benighting and devouring."

Milton's passage on the Plague is noble :—

> " As when the potent rod
> Of Amram's son, in Egypt's evil day,

> Wav'd round the coast, up call'd a pitchy cloud
> Of locusts, warping on the eastern wind,
> That o'er the realm of impious Pharaoh hung
> Like night, and darkn'd all the land of Nile."

And Heber too rises to the theme :—

> " The dreadful wand, whose godlike sway
> Could lure the locust from her airy way,
> With reptile war assail their proud abodes,
> And mar the giant pomps of Egypt's gods."

But I do not know where else in poetry to look for an adequate reference to this terrific phenomenon of the locust, the little insect which the Arabs grind up to make flour for cakes, yet compared with which the devastating armies of man are benevolent agencies.

Marlowe's little reference to the girl's traps to catch locusts is pretty, and, when we think what the locusts really are, pathetic too :—

> " A country virgin keeping of a vine,
> Who did of hollow bulrushes combine
> Snares for the stubble-loving grasshopper."

I have myself followed both army and insect.

Where the army had passed, the villages were empty shells, the green crops had been cut down lest they should ripen, the melon-fields hacked to bits lest they should bear fruit, the wells befouled with the carcasses of dead beasts. Fire had been there, and the fury of swords. And yet there was greenness left, and, though of a poor sort, gleanings for animals. The injury done was not intolerable; the land was habitable.

In the other case there had been neither brand nor blade, and no malevolence. And yet there was nothing left, neither for the camel searching the tops of the mimosas nor for the mule sniffing for herbage between the stones on the ground. The earth was burned close. The bushes were more bare than in mid-winter. The only well we found stank to the skies with a fathom's depth of dead locusts. It was the difference between discomfort and starvation, mischief and ruin. And Joel says, " The appearance of them is as the appearance of horses ; and as horsemen, so shall they run." The translation probably is in error, and, I should think, should read, "and their coming is as the coming of horses and as horsemen, so shall they run."

Their passing was like the rush of infinite cavalry at a distance, the air all rattling with harness and glinting with sparks of silver and steel and scarlet. "And the Lord shall utter His voice before His army"—that voice of the whirlwind with which Jehovah spoke in the desert to Moses ; the voice of nature real, earnest, indisputable, and authentic.

CHAPTER VI.

LUCIFERS AND THE POETS.

> " They made her a grave too cold and damp
>   For a soul so warm and true,
>   And she's gone to the Lake of the Dismal Swamp
>   Where, all night long, by a firefly lamp
>   She paddles her white canoe."—*Moore.*

ONE of the most satisfying and impressive facts in nature is, I think, the illumination by phosphorescent creatures of the deep sea.   Popularly, the ocean depths stand as a synonym for more than sepulchral darkness, and in legends and poetry some incidental lustre is imagined in order to make the profundity fit for the habitation of the sea-folk.   So their cavern-palaces are lit up by gems.

Sometimes, as in Keats, where the hero follows the curves of the shore in his sub-aqueous excursions, and so keeps in the comparative shallows, the water is of course sufficiently translucent to afford the Nereids' grottos and the mermaids' haunts a soft dim light of deep sea-green.   But the real abysses of oceans are not the scenes of adventure either in legend or in poetry, for these concern themselves only with the smaller seas and with bays and straits and rocky coasts. In the days when legends were making, the mid-ocean lay outside the sphere of song-smiths' knowledge, and almost of their speculation.   It had its monarch, and he his court; but their apparitions and apocalypses occurred only in waters within the mariners' ken, and, as a rule, at such distance from the shore that men standing on the cliff could point out with the arm to the haunts of the water-people, the Mediterranean's bay-indented coasts, resounding Scandinavian fiords, or labyrinths of Grecian archipelagos. But the veritable hollows of the sea, its more prodigious depths, were a mystery that lay below even the soundings of fancy; so they remained a blank of intense darkness, these inmost recesses of ocean.

Now, however, we have learned something of the truth; and one of the secrets that science has discovered is certainly as fascinating as anything even in astronomy. For it is known that (in many places at any rate) the ocean depths are not dark at all, but illuminated perpetually by myriads of living lamps; and what is true of the known may possibly be true of the unknown.

Down, down, down you may go till the sea is nearly at freezing-point, the pressure two or three tons to the square inch, and some four miles of water rolls between you and the keels of over-passing ships; and yet all through the descent, and when you can get no deeper, you will find that the sea is still faintly luminous, and that the creatures about you live in twilight, seeing each other come and go.

Myriads of tiny things, living closer than blades of grass in a meadow, than leaves on forest trees, illustrate each their own specks of space, till the whole territory is luminous, and speck by speck, and inch by inch, cubic miles of sea glow incandescent. Among these atoms, these motes of light, move other and larger bodies, ranging from the great cyanea with sparkling tentacles that trail fifty yards behind it, to the monads invisible to the naked eye, yet which if a tumbler is filled with them show light enough to read by. One jellyfish, the largest invertebrate animal ever seen (it weighed two tons) was stranded near Bombay, and for several nights its body was visible half a mile away as a shapeless luminous mass. Darwin tells us how his nets sparkled brightly the night after they had been used to drag the deeps. A dead aurelia put into a quart of water made the whole so luminous that cards could be comfortably played by its light. One night in a phosphorescent sea a ship's crew beheld a water-spout as a moving pillar of solid fire. Sir Wyville Thompson dredged up from 600 fathoms mud that was like pure gold. A haul of starfishes off the Shetlands flashed like brilliants in the nets. Boats have rowed for miles over green lambent flames, a forest of the luminous sea-pens. On the Patagonian coast, after a storm, virgularia lay heaped so high on the beach that at a distance it seemed as if the watch-fires of an army were burning all along the shore. Off Somerstown for a week together some disturbance drove the creatures of the depths to the surface, and the crew of a vessel lying there read at the portholes all night through. Holden tells us of wondrous displays in the Southern Seas—how, "drifting over coral reefs, he saw the bottom studded with gleaming gems, yellow and purple gorgonias bathed in soft lights, which, when lifted to the surface, illuminated all about them with a mild radiance; flashes of light came and went, appearing again in the distant depths like spectres; the silver sand, turning over on the oar that disturbed it, flashing with sparks of living light; processions went winding by, breaking up, and reforming in aggregations of light, nebulæ of breathing stars." Humbolt, passing through a zone of pyrosomas in the Gulf Stream, could distinguish by their light far down below the surface the forms of dolphins and other fishes, thrown up in strong relief against the gleaming myriads. Naturalists have again and again written down the descriptions of these creatures by their own light. Off the Mauritius amazing scenes have been beheld, especially after violent storms. Moseley captured a pyrosoma four feet long, which when touched glowed like metal at a white heat, and flashed for hours afterwards in brilliant colours, as if chemicals were being thrown on a molten surface.

But enough of facts. These few suffice, however, to show

how various, how important in combination, the light-givers of the ocean are. Yet even they, the radiant bodies in motion, do not suffice for nature. For she has made many, if not all, the submarine corals phosphorescent, and much of the deep-sea sand is filtrated with minute organisms that are on occasion, or always, luminous. As if this were not enough —that she should provide the fishes with incandescent walls and floors, and should set afloat in all directions moving lamps—nature has affixed to the rock-edges " where young lampyris waves his plumes of gold," deep under water, and to the dangerous cliffs which the mariner-fishes might strike on if not for danger-signals, actual lighthouses, and with revolving lights too ! These are some of the anemones, which, as they throw out and withdraw their tentacles, alternately show and extinguish a beacon on the brink of projecting ledges or the entrances to gloomy caverns. Others, again, like the pholas, are veritable lightships, having lustrous bodies enclosed in shells, and steering their way along the edges of the submarine cliffs as if warning off the incautious navigator.

Is not all this wonderful? But surely it surpasses all— the regular electric light "laid down" as it were, the phosphorescent walls and so forth, the lighthouses and the lightships—that there should be races of fishes who carry bull's-eye lanterns about with them. For what else is the ipnops which we read of in the *Challenger's* voyages, whose two eyes throw out blazing rays before it as it swims? or what else are the score of species which have their bull's-eyes on their flanks, and, like a hansom with its lamp on either side, shoot along by the light of the lights they carry?

What a delightful range of subject and metaphor all this, had it been known, would have offered to the poets of pre-Tennysonian days ! How many old ideas would have had to be dismissed ! how many more beautiful might have taken their places !

As it was, the poets' only glimpses of nature's ingenuity in lightening the darkness was "the earth-wandering star" (the glow-worm), and the "traveller's friend" (the firefly). Yet each in its small sphere is an exquisite touch. Indeed, among the prettier devices of nature must take high rank the "impassioned light" (as Darwin calls it) which the glow-worm holds out to tell her winged Leander the way across pathless wastes to her bower. It is true the Leander himself sometimes shows a faint sympathy of incandescence; but such luminosity is not more regular than beards on women's chins. The privilege of shining belongs to the fair sex alone.

> " When evening closes nature's eye,
>   The glow-worm lights her little spark
>   To captivate her favourite fly,
>   And tempt the rover in the dark.
>
>   Conducted by a sweeter star
>   Than all that deck the fields above,
>   He fondly hastens from afar
>   To soothe her solitude and love."—*Montgomery.*

> " Innocent as is the light
>   The glow-worm hangs out to allure
>   Her mate to her green bower at night."

No one who has watched the insect on its grassy bank can have failed to notice that its light seems green.

> " The fireflies played around the pillared stems,
>   And bore about their lanterns of green light,
>   Advancing and receding while the eye
>   Measured by them the depth of sylvan gloom."—*Faber.*

Some poets, however, call it "silvery," which it surely never is; others "golden," which it only sometimes is; and one, Eliza Cook,[1] has "*azure ray.*"

Shelley's "like a glow-worm golden, In a dell of dew" is, like everything of Shelley's, exquisite. His admiration of night and the natural features of night is always eminently

---

[1] Did she get the idea from Charlotte Smith's "sapphire beam" of the firefly?

noteworthy. His poems are full of delightful moths, and he is tender to the owl, "sad Aziola." Both glow-worm and firefly sparkle throughout his verse. And, indeed, who knew the twilight and the starlit hours so well as he? He has a charming fancy about the glow-worm, that it lives in lilies, and that the petals fold over it to keep it from the dew, which otherwise might quench its tiny spark. In the "Sensitive Plant," the flowers drooping at evening "fall into pavilions white, purple, and blue, To roof the glow-worm from the evening dew"; and, again, in the "Witch of Atlas,"

> "A green and glowing light, like that which drops
> From folded lilies in which glow-worms dwell,
> When earth o'er her face night's mantle wraps."

I do not understand the "dropping" of the light—the poet repeats it twice elsewhere—for the glow-worm's light always strikes me as being constant in place even though fluctuating in degree. It certainly is not, as many poets describe it, "glancing," or "glittering," nor can Cunningham's moral commend itself from any point of view, whether real or imaginative.

> "How bright the little insects blaze
> Where willows shade the way,
> As proud as if their painted rays
> Could emulate the day.

> "'Tis thus the pigmy sons of pow'r
> Advance their vain parade,
> Thus glitter in the darken'd hour,
> And like the glow-worms fade."

The "modest" and "humble" glow-worm are both better, and its light is best described in such of the poets' phrases as "lustre mild," "pale lustre," "soft green light," or "harmless ray."

Barry Cornwall has a sympathetic verse :—

> "Night's shining servant! Pretty star of earth
> I ask not why thy lamp does ever burn ;
> Perhaps it is thy very life—thy mind ;

> And thou, if robbèd of that strange right of birth,
> Might be no more than man—when death doth turn
> His beauty into darkness, cold and blind."

There is one idea, sufficiently obvious, which might occur to any imaginative intelligence as he watches the insect after nightfall "move with green radiance through the grass, An emerald of light;" but yet it is one which, remembering not only the gentle circumstances of this little creature's effulgence, its wingless groundling state, and the dangers which its far-seen affection attracts, but admiring also its curious beauty—the little "fairy-lamp" of elfin revel, the wee pale glimmer in the green—I, for one, would rather forego. What if the creature is only a grub-like obscurity by day, and comes forth in the dark to shine and to allure? The analogy ends there. For half of the insect world our day is their night, while as for the glow-worm herself, her taper is meant *for one alone;* and just as among the Ana the wife hangs up her wings after marriage, to show that she has found a home and final abiding-place at last, so the glow-worm, once mated, pales her useless fire and forgets that once she shone. But not so the poets, who tatter the idea to shreds.

The following represents compendiously the aspects which it assumes :—

> "Thus oft we see a glow-worm gay
> At large her fiery tail display,
> 　　Encouraged by the dark ;
> And yet the sullen thing all day
> Snug in the lonely thicket lay,
> 　　And hid the native spark."—*Fenton.*

> "But thou, with spirit frail and light,
> 　Wilt shine awhile and pass away,
> As glow-worms sparkle through the night,
> 　But dare not stand the test of day."—*Byron.*

> "Confiding glow-worms, 'tis a night
> Propitious to your earth-born light,

> But, where the scattered stars are seen
> In hazy straits the clouds between,
> Each, in his station twinkling not,
> Seems changed into a pallid spot."—*Wordsworth.*

> "Warm on her mossy couch the radiant worm,
> Guards from cold dews her love-illumin'd form ;
> From leaf to leaf conducts the virgin light,
> Star of the earth, and diamond of the night."—*Darwin.*

Its transcience, in that melancholy tendency of the poets to see vanity in everything, is constantly adverted to ; its light is "feeble," "fitful," a "soon-quenched" spark. In the same vein, too, are the following :—

> "The man who first upon the ground
> A glow-worm spied, supposing he had found
> A moving diamond, a breathing stone ;
> For life it had, and like those jewels shone ;
> He held it dear ; till by the springing day
> Informed, he threw the worthless worm away."—*Waller.*

> "If on some balmy breathing night of Spring
> The happy child to whom the world is new
> Pursues the evening moth of mealy wing,
> Or from some heath-flower beats the sparkling dew,
> He sees before his inexperienced eyes
> The brilliant glow-worm like a meteor shine
> On the turf bank, ...
>
> .   .   .   .
> Yet with the morning shudders to behold
> His lucid treasure, rayless as the dust."—*Charlotte Smith.*

How the nightingale and the glow-worm made sad acquaintance is a fable that is well known. So Marvel's opening stanzas of "The Mower" are all the more delightful :—

> "Ye living lamps, by whose dear light
> The nightingale does sit so late,
> And studying all the Summer night
> Her matchless song does meditate.
>
> Ye country comets, that portend
> No war, nor prince's funeral,
> Shining with no higher end
> Than to presage the grass's fall.

> Ye glow-worms, whose officious flame
> To wand'ring mowers shows the way,
> That in the night have lost their aim
> And after foolish fires do stray."

But more generally accepted is the version that points the moral of the danger of lamp-carrying when the enemy is abroad in the night :—

> " The glow-worm must shine, though the light which it shows
> But guides the attack of its wandering foes."

And so it happened in the fable, for the bird was looking on while the insect crept in conspicuous phosphorescence in the herbage below :—

> " Nor spares, enamoured of his radiant form,
> The hungry nightingale the glowing worm ;
> Who with bright lamp alarms the midnight hour,
> Climbs the green stem, and slays the sleeping flower."—*Darwin.*

Moore, too, relates the episode, drawing the moral from the fable :—

> " The prudent nymph, whose cheeks disclose
> The lily and the blushing rose,
> From public view her charms will screen,
> And rarely in the crowd be seen ;
> This simple truth shall keep her wise,
> The fairest fruits attract the flies.
> One night a glow-worm, proud and vain,
> Contemplating her glittering train,
> Cried, ' Sure there never was in nature
> So elegant, so fine a creature.'
> She spoke : attentive, on a spray,
> A nightingale forbore his lay,
> He saw the shining morsel near,
> And flew, directed by the glare ;
> Awhile he gazed with sober look,
> And thus the trembling prey bespoke :
> ' Deluded fool ! with pride elate,
> Know 'tis thy beauty brings thy fate ;
> Less dazzling, long thou might'st have lain
> Unheeded on the velvet plain,
> Pride soon or late degraded mourns,
> And beauty wrecks whom she adorns.' "

But Cowper represents the nightingale as won over to mercy by the glow-worm's appeal :—

> "'Did you admire my lamp,' quoth he,
> 'As much as I your minstrelsy,
> You would abhor to do me wrong,
> As much as I to spoil your song ;
> For 'twas the self-same Power Divine
> 'Taught you to sing and me to shine ;
> That you with music, I with light,
> Might beautify and cheer the night.
> The songster heard his short oration,
> And, warbling out his approbation,
> Released him."

Wordsworth sees it offering " nightly sacrifice " to the skies ; and Savage is excellent in his glow-worm's "glimmering through the night, And scattering like hope through fear a doubtful light." For it often seems, when watching "in the grass-green haze the glow-worm's living light," that they are trembling every minute on the point of going out.

With the necessary differences on account of its being a foreigner, and therefore comparatively unfamiliar, the firefly is virtually a repetition of the glow-worm—but on wings.

Columbus finds (in Rogers's verse), the firefly "spangling the locks of many a maid ;" and, again, the poet familiar with "the shining race in Tuscan groves " sees the Roman Floretta "spangling her hair with stars." In Moore they frequently recur, once in connection with "the favourite tree of that luxurious bird which lights up the chambers of its nest with fireflies ;" and again as used in palace illumination—

> "The chambers were supplied with light
>   By many strange but safe devices,
> Large fireflies, such as shine at night
>   Among the Orient's flowers and spices."

Once upon a time, when I was at large in the United States, I took that wonderful trip from San Francisco to St. Louis by the Texas and Southern Pacific railways. I have notable memories of that journey, and many touches of nature remain on the mind—the huge sea-lions in the Pacific below the Cliff House at San Francisco, clambering on their rocks of refuge, sprawling, scuffling, splashing; the owl-lands of New Mexico, where bird and snake and ground-squirrel live together; the cities of the prairie dogs; the bee-ranches; the leagues of yucca in full flower; the wonders of the cactus.

But above them all stands out the firefly country of Texas and Arkansas, where the land is all swamp, and the old haggard trees, tapestried with ragged moss, wade ankle-deep in brown stagnant water. The forest glades are long pools, and wherever a vista opens there is a thin bayou stretching away between aisles of sombre moss-ragged trunks. There is a strange antediluvian gloom about the place—this forest standing in a lagoon.

The world was something like this when the Deluge was subsiding. Uttermost silence abides here; except when a turtle stirs in the mud, or a water-snake makes a ripple on the dull pools. Sunlight! Not a ray of light ever pierces to the roots of the trees. But at sunset, when the orb goes down rosy-red behind the water-logged trees, and their trunks stand out black against the glaring sky, and the pools about their feet take strange tints of copper and purpled bronze—what a sight it is!

The railway pierces an avenue straight as an arrow for miles and miles through the belt of forest. On either side along the track lie ditches filled with water. And at sunset the ditches seem all filled with blood, and the sky seen away in the distance underneath the trees hangs like a furious crimson curtain.

And as soon as the sun begins to set the awesome forest-swamps awake, the sluggish waters lap and mumble upon the snags as the creatures that live in them arouse themselves, and out from the rotten heaps comes the frog, and out from their dormitories on the boughs the katy-did. "Yank," falters the one. "Katy *did,*" replies the other. And thereupon the spell is broken.

As if by some talisman, the charm of the enchanted sleep is snapped, and in an instant there grows upon the air such a volume of sound as beggars language. The fatal word of the fairy-story is spoken, and on the instant the tempests rage. Every inch of swamp has its "mud-compeller" in full song, every inch of forest overhead its katy-did in ceaseless clamour, assailing and pursuing the flying train. It sounds as if the whole country were screeching and jeering at the carriages as they fly by. Stop thief! stop thief! Katy-*did!* Yank! Yank!

No wonder Greece thought of the frog as it did! We in England know nothing of the "damnable iteration" of its multitudinous rioting. But Arkansas does. Like the stars past counting, and the wrinkles on the sea beyond arithmetic, the pen shrinks from description of the din. It is like launching on "the tenth wave" for an infinite natation upon cycles of floods. The subject is impossible; for the air vibrates, throbs, seems bursting, and ready to crack with, the soulless metallic babel of the frogs. Yank! Yank! Yank! And yet through it, over it, under it, all round it, shrills supreme the strident cry of the cicadas, without number and without shame, countless katy-dids swearing their confidences to all the reluctant stars. It is no use trying to talk in the carriage; all you hear is "Yank! Katy-did! Yank! Yank! Katy-*did.*"

It was enough to make Arkansas a memory for ever. And yet, whenever I remember that most strange railway journey, it is neither swamp, nor frog, nor katy-did that first recurs to me, *but the fireflies.* The black night, with its glistening pools, its interminable clamour of brazen-throated batrachians and tin-lunged cicadas are forgotten, while recollection conjures up afresh that miracle of the lantern-bearing myriads.

As far as the eye could reach into the water-logged forests, and up in the air among the invisible branches of invisible trees, flickered in inextricable bewilderment hosts of fire-flies. The air was thick with them. And such flighty fitful creatures as exasperated the intelligence, perpetually striking matches as if to look for something, and then blowing them out again, flick! flick! flick! in countless millions; and all apparently desperate of any other purpose but to confound confusion. The frogs sounded as the sand of the shore for multitudes, and there were at least two katy-dids for every frog; and yet multiply the frogs by the katy-dids and you would not get the total of the fireflies. It seemed as if each "yank" and "katy-did" struck out fireflies just as flint and steel strike flashes, or as if the recriminations of batrachian and insect caught fire as they flew and peopled the inflammable air with phosphorescent points of flame; a battery of din perpetually grinding out showers of sparks.

In the poets, however, it flits, "a bright earth-wandering star," "tall majestic trees between;" it plays "around the pillared stems," sparkles "in deep cedars" and "through the brakes," lights up "forest bowers." But the "emerald" light no longer serves, for "the fireflies' lanterns of green light" now becomes as inaccurate as before it was correct. Shelley repeats his glow-worm fancy in the lines—

> " Carved lamps and chalices which shone,
>   In their own golden beams, each like a flower,
>   Out of whose depth a firefly shakes his light,
>   Under a cypress in a starless night."

In another poem he has both insects together—a license of geography that entomology scarcely, I fancy, authorises.

> " Day had awakened all things that be—
>   The lark, the thrush, and the swallow free;
>   .   .   .   .
>   Fireflies were quenched on the dewy corn,
>   Glow-worms went out on the river's brim,
>   Like lamps which a student forgets to trim;

> The beetle forgot to wind his horn,
> The crickets were still in the meadow and hill;
> Like a flock of rooks at a farmer's gun,
> Night's dreams and terrors, every one,
> Fled."

Without much industry the poets discover that (like the glow-worm) the firefly is only luminous in the dark :—

> " This morning when the earth and sky
>   Were burning with the blush of spring,
>   I saw thee not, thou humble fly,[1]
>   Nor thought upon thy glowing wing."

So, again, Rogers, who, to do that poet justice, had evidently (from the last two lines) really noticed a firefly :—

> " There is an insect, that, when evening comes,
>   Small though he be, and scarce distinguishable,
>   Like evening clad in soberest livery,
>   Unsheaths his wings, and through the woods and glades
>   Scatters a marvellous splendour.   On he wheels,
>   Blazing by fits as from excess of joy,
>   Each gush of light a gush of ecstasy."

But perhaps it is worth noting as the experience of one who has travelled that the firefly does *not* show the traveller the way.   Of course in imagination it seems obvious that it must ; but this is just one of those occasions when imagination is useless.   For the sober fact is very much to the contrary.

To the traveller the firefly is an unmitigated nuisance. Seen at first, it pleases ; the spectacle is engaging.   But when the myriads break from their resting-places as night

---

[1] Moore's moral is this—
> " But now the skies have lost their hue,
>   And sunny lights no longer play,
>   I see thee, and I bless thee too
>   For sparkling o'er the dreary way.
>   Oh ! let me hope that thus for me,
>   When life and love shall lose their bloom,
>   Some milder joys may come like thee,
>   To light, if not to warm, the gloom."

draws on, and flicker up and down in mazy multitudes, the result is utterly baffling and bewildering.

And herein (of which I believe I am now the first true exponent) lies the secret of the legend of the will-o'-the-wisp and the *ignis fatuus* generally.

Naturally enough, a belated traveller looks upon a light as a friend. There is a firefly in his path. Therefore it is his friend. By-and-by he cannot see his path. But there is another firefly. So he thinks, That is "another friend." Alas! it is anything but that, and will lead whomsoever follows it into uttermost swamp and jungle —the natural habitats of the creature. This is no effort of imagination.

In the years that are gone I was riding in Sind, the second march beyond Jacobabad, and for several miles at midnight my way lay through woodland. A straight road, at least twelve feet wide, had been driven through it for our artillery and waggons to go by, and it was as flat and soft as tan. Had the night been pitch-dark I could have cantered the whole way. But, as it happened, the place was ablaze with fireflies, and *I could not see an inch before me.* Wherever there was a natural glade or break in the vegetation, to right or left, there at once opened out a glittering vista of these dancing, flickering creatures. You could see, apparently, a road, lamp-lighted, straight before you. But turn round. It was the same behind you. There was a straight road. Look to the right or the left. There, too, stretching away into a hazy confusion of twinkling points, were straight roads too. Stupidly mazed, I once got my horse's head round; but the beast was more sensible than I, and told me the right direction, and I listened to it.

So to speak of fireflies being a help to the traveller is utter nonsense; they are the most provoking distracting mischief-makers possible. Leave a path alone, and even on the darkest night a sober man may learn to distinguish it.   But light up tens of millions of fireflies in the bushes on either side, in all the openings in the undergrowth, and the path, even though twelve feet wide and straight as an arrow, is nowhere.   It is one of the frauds of poetry, but, all the same, one of the charms of nature.

CHAPTER VII.

DEBORAH: "THE HONEY-BEE."

SUPREME among the poets' insects is the honey-bee. Indeed, it occupies quite as large a space in verse as either the dog or the nightingale, the dolphin or the rose, among the bards' beasts, birds, "fishes," and flowers.

For Deborah—the name "being translated means the honey-bee"—abounds, even on the surface of its small brown individuality with morals and metaphors and similes. Chief of these are its diligence, as "the busy bee;" its "sweet alchemy," as the honey-gatherer; its "skilful architecture," as the artist "that builds its golden house downward;" its sense of discipline, as a citizen under a constitutional government :—

> "Creatures that by a rule of nature teach
> The art of order to a peopled kingdom."

Besides these, the insect has many engaging traits and habits, each of which suffices for endless illustration of human analogies; and so, through all the poets' "immemorial" lines, is heard "the murmuring of innumerable bees."

Yet of all bee-facts and bee-fictions nothing attracts the poetic fancy more than the honey-gatherer's association with flowers.

In April, when the sallow is abloom with golden catkins, faintly fragrant, like Australian wattles, the first of the honey-

folk are abroad, and the voices of them busy at their work
fill the streamside with a sweet premonition of summer's
June, and the dreamy murmurings of bee-beleaguered
hives.

> " This renovating season, too, calls forth
>   The humming tribes, for now the million leaves
>   And downy flowers, or river-loving palms,
>   Afford material for the curious cell ;
>   And oft, e'en in this chill ambiguous month,
>   The labourer returns with loaded thighs."—*Grahame.*

And so all summer and autumn through, with lavish choice
of flowers, to spare November, when the ivy—" my yellow
bees in the ivy-bloom "—tempts here and there the last
labourers of the hive.

The foxglove is a favourite bee haunt.

> " The bee's low tune in the foxglove's bell."

> " Where the bee's deep music swells
>   From the trembling foxglove bells."

And no one who has passed along the copse-side, or sat
on the bank among the foxgloves, can have missed that
deep hum of a full content as the bumble-bee scrambled
from flower to flower, leaving them all " wagging their sweet
heads," as it suddenly made up its mind for home, and
flew off on heavy wings, drawing out behind it as it went
a tapering note.

> " To keep her slender fingers from the Sun
>   Pan through the pastures oftentimes hath run,
>   To pluck the speckled foxgloves from their stem,
>   And on her fingers neatly placèd them."

It is a delightful flower, the pet of legend, and a grace to
every spot.

> " Upon the thistle-tops and heather bells."

I have myself seen bumble-bees in their velvet jerkins
belted with yellow, probing the thistle heads ; but I often
wondered why. There is a scent about the flowers, it is

true, but the honey, such as there is, must be very hard
to find, and is thin when found.

No wonder the poet says—

> " Plying his earnest task,
>   The bee in the bells of thyme."

What a musical line it is !   And whether thyme has bells
or not, the bees love the flower, and I remember many a
time and oft, when I lived in White of Selborne's country,
lying on the bank crushing the wild thyme where it grew,
listening all the time I read to the murmur of a multitude
of bees, with the air about me aromatic with the Attic
herb.

> " And from the knotted flowers of thyme
>   When the woodland banks are deck't,
>   See the bee his load collect ;
>   Mark him turn the petals by,
>   Gold-dust gathering on his thigh,
>   As full many a hum he heaves,
>   While he pats th' intruding leaves
>   Lost in many a heedless spring,
>   Then wearing home on weary wing."—*Clare.*

So too, as often as I think of my Kentish heath, "gleaming in purple and gold "—the purple of heather and the
gold of the whins—I remember the flower and the insect
together, and that device of chivalry, the bee on the tuft of
heather.

I do not know them, "humming among the cowslips," or
"busy in the golden broom ;" though, as Bloomfield and
Clare are beautifully observant, I expect the bees know
both.   But who can have missed—

> " In immemorial elms,
>   The murmur of innumerable bees ;"

or in the garden, when the currant-bushes are in flower, the
myriad voices of the early swarms; or the space-pervading
drowsiness of the "honey-flies," on the new-blown heath ; or
" haunt of birds and bees, the purple moorland ;" or "the

sycamore musical with bees ; " or the bean-field, heavy with evening scent ; or the " honey'd lime all murmurous with bees ; " or the orchard trees, apple and pear and plum, beset by marauding hosts of happy plunderers ; or busy about Keats' " globes of clover," so that the whole field is humming, and the restless butterfly can hardly find a blossom to sip at undisturbed ?

> " Crowds of bees are *giddy with clover*,
> Crowds of grasshoppers skip at our feet."

Jean Ingelow puts her bees in the white trefoil—

> " Murmuring in the milk-white bloom
> As babes will sigh for deep content."

Others in " the trumpets of the eglantine," and, let antiquaries say what they choose, Milton in the lines—

> " Sweet briar or the vine
> Or the twisted eglantine,"

meant the honeysuckle, the

> " Honeysuckle full of clear bee-wine "

of Keats' delicious verse ; and so too mean the other poets that make the same slip. Sometimes they call it woodbine, " suckling the bee with honey and the moth " (Hurdis), and sometimes eglantine, but they always mean the same sweet plant, the honeysuckle, emblem of fidelity in love, and beloved by bees.

To clear the way before coming to the poets' Deborah in her home-life, and as explanatory of their allusions, I must return for a paragraph or two to the old subject of the natural history of the real bee. The shorter my reference the better.

In Spring, then, the community consists of only one fertile female bee, " the queen," and a staff of workers, " the neuters," and as soon as ever the warmer weather enables

the half-dormant creatures to move about, the latter commence their toil.　The hive is thoroughly cleaned up and
set to rights after the wear and tear of a winter's habitation,
and as soon as the flowers begin to bloom, and buds to
show, they begin the work of collecting honey to refill the
cells which have been emptied during the cold weather;
pollen for mixing with honey and making "bee-bread" to
feed the coming grubs with; and "propolis," the resinous
exudation of buds, for use as glue, caulking, and varnish for
repairing, fastening up, and finishing off the cells.　These
cells are built of "wax" (a substance which the bees do
not collect, but which exudes from their bodies, an involun-
tary secretion), and are made of four sizes, for holding the
grubs of "neuters," "drones" (or male-bees), and "queens"
respectively, and as jars for honey and bee-bread.　They
are as a rule six-sided.　But this form results not from
any hexagonal design on the part of the bees, but from the
manner of their building, and the pressure upon one another
of six cylinders being built simultaneously.　The exceptions
are the cells upon the edges of the comb, or isolated cells,
for these, having no such pressure on their sides, remain
cylindrical.　In each cell the queen deposits an egg, which
hatches into a grub; this again turns into a chrysalis, and
from the chrysalis—in about three weeks' time, more or
less, from the laying of the egg—there emerges a bee.
According to the size of the cell in which the egg was
laid, and the kind of food upon which the grub was fed,
the bee that emerges is either a neuter or male.　In a
few cells, which are often shaped like acorns and are much
larger than the rest, are laid eggs of which the grubs are fed
entirely upon a special "royal" jelly instead of ordinary
bee-bread, and these grubs turn into queens, or would do
so if the old queen did not, on discovering that another
female had come forth, go round and kill the rest.　The
young queen that has already issued is carefully protected

by the workers from the old one, and she soon flies abroad upon her wedding tour, and then returns to the hive. Meanwhile, the old queen has been very restless, and after the young queen's return she leaves the hive accompanied by a mob of neuters, to found a new colony. This is a "swarm." The new queen very soon commences her duty of laying eggs—at the rate of a hundred an hour, and laying in the course of a year about two hundred thousand— and the males, being no longer required, are massacred by the workers. And so the Summer and Autumn pass, the queen assiduously laying eggs; the neuters as diligently rearing the young, piling up honey, and dying. And so to Winter, when the hive becomes partially dormant until Spring comes round again.

Such in briefest outline is hive-life, and absolutely bare as it is, sufficiently interesting. But the actual details which would fill it out in a completer chronicle (but with which this article has nothing to do, except where the poets refer to them) are such that fancy and legend cannot exaggerate their wonders, and severest science is unable to refrain from admiration and amazement. These details, however, belong to the natural history of bees, and not to the unnatural history of the poets' insect.

Many poets give sketches of the life of the honey-gatherers :—

> " All hands employ'd, the royal work grows warm
>   Like lab'ring bees on a long summer's day,
>   Some sound the trumpet for the rest to swarm
>   And some on bells of tasted lilies play ;
>   With glewy wax some new foundations lay
>   Of virgin comb, which from the roof are hung,
>   Or tend the sick, or educate the young."—*Dryden.*

> " As in the wingèd commonwealth of Bees
>   (Whose careful summer-providence foresees
>   Th' approaching fruitless winter, which denies
>   The crown of labour) some with laden thighs

> Take charge to bear their waxy burthens home,
> Others receive the welcome load ; and some
> Dispose the wax, others the plot contrive ;
> Some build the curious comb, some guard the hive,
> The armed Centinels ; others distreine
> The pure honey from the wax, some traine
> And discipline the young, while others drive
> The sluggish Drones from their deservèd hive."—*Quarles.*

No other insect avails itself so cordially of human co-operation. The silkworm, for instance, is delicate, fastidious, and whimsical. The cochineal is obstinate, implacable, and barbarous in its diet. But the honey-bee asks for nothing from its cultivator but cleanliness, protection from extremes of temperature, and occasional assistance in food and drink. Once put up a hive in a sheltered spot, near water if possible, but in any case within easy flight of pasturage, and the combs will be kept as regularly filled with exemplary honey as if the insects had signed contracts for the work, and the queen-bee herself had gone guarantee for their fulfilment. It is not even necessary to pamper Deborah's sweet taste, since the insect knows far better than man can tell it the flowers which pay best, and can give it in wax and honey the fullest returns for its labour. In the waste hedge-strips of the field, the purple corners of spinney and copse, it finds the true bee pastures, ditches filled with teazel, burr-dock, and dandelion, wild cabbage, and wild turnips, banks overgrown with trefoils, rest-harrow, thistle, and nettle.

Gardens brilliant with all that gardeners delight in will not cause the bee to stop when it is on its way to a field of sainfoin, nor can all the triumphs of horticulture—let the colours be never so bright, the perfumes never so dainty—tempt it to alight when on the wing towards a hedge that is overgrown with honeysuckle, or a lane entangled in clematis. Nature herself has taken great care of the bee :—

" I know where the meadow-sweets exhale
And the white valerians load the gale,
I know the spot the bees love best."

" And on the bosom of the hills,
    Wooing the bees, the modest heather
Waves to the wind its hardy bells."

" From flower to flower the busy bee
        With anxious labour flies,
Alike from scents which give distaste
By fancy as disgusting placed,
        Repletes his useful thighs.
Nor does his vicious taste prefer
The fopling of some gay parterre,
        The mimicry of Art,
But where the meadow-violet dwells,
Nature replenishing his bells,
        Does ampler store impart."

The centre figure of the home-life of bees is of course the queen-bee, one of the chiefest wonders of the animated world. Volumes of facts have been published of the almost incredible life-history of this insect, of its creation artificially by the drudges of the hive, of its terrible life. A truly awful insect; a made-up monster as it were with a body complete in all its parts, but no soul, no heart. Of the idolatrous devotion of the hive to this tyrant, literally of their own construction, endless fanciful prose has been written, but in verse it finds scarcely more than such passing notice as when the Midsummer Fairies—

" Huddle in a heap and trembling stand
All round Titania, like the queen-bee's band."

Then too the drones, the queen's paramours, "vex'd and murmuring," says Keats, "like any drone shut out from the fair bee-queen," whom the neuters pamper and caress, until one day when, the population being equal to the space, they fall upon the honey-fattened stingless throng and sting them all to death !

> " As they sat,
> A bee alighted on a clover tuft
> Ceased for an instant its laborious hum,
> And peer'd in petals of the purple flowers
> With busy pleasure.   ' There,' said Montague,
> ' Yon little insect, wiser than mankind,
> Might teach the world a lesson that it needs.'
> ' What ! in its ruthless murder of the drones
> And pampering of a fat luxurious queen ? '
> ' Not so, but in its love of daily toil,
> A toil unselfish.   In the social hive
> One labours for the whole community,
> And the community for every one.
> Toil is their joy.' "

> " Without a soul ! unless the bees have souls !
> These yield a blind obedience to their chief,
> And feed and swaddle it and make it fat,
> And toil and moil until th' appointed hour
> When in hot swoop they fall upon the drones,
> And kill the fluttering fathers of the state
> Or may be choose another sovereign."—*Mackay.*

For a long time it was supposed that the sovereign
of the hive was a male, "the king;" and I think this
sentence from Pliny is worth quoting, "Hath the king
of the bees alone no sting, and is he armed only with
majesty? or hath nature bestowed on him a sting and
yet denied him only the use thereof? For certain it is,
that this great commander over the rest does nothing
with his sting, and yet a wonder it is to see how they all
readily obey him." Virgil has a king-bee, and he describes
how his subjects defend their monarch :—

> " Onward they troop, and brandishing their wings,
> Fit their fierce claws and point their poison'd stings,
> Throng to the imperial tent, their king surround,
> Provoke the foe, and loud defiance sound."

That the queen-bee has a sting is abundantly proved
every year by her using that weapon to kill her possible
rivals, but it is one of the curiosities of bee-life that,
however irritated (and her subjects treat her with the

scantest possible ceremony on occasion), she never uses her sting against them. " Majestate tantum " (by majesty alone), was therefore a favourite motto of principalities and powers. Among others it was the motto on the bee-crest of Louis XII. of France, "Père du Peuple." But the fancy was surely stretched to the snapping-point when the " Maid " of Orleans was given a queen-bee for her impresa.

When the hive is over-crowded, the conservative party of the community secedes from the parent-stock, taking the old queen with it, to found a new colony. As soon as the owner or a friendly neighbour hears " the murmuring sound of the swarming bees," he comes out—the practice is still almost universal among the countryfolk—and striking a key on a pan, or jingling the fire-irons together, "calls" the swarm.

> " So on a Sabbath morn
> Cloudy and calm, with not one sunny gleam
> To lure them forth, I've seen a num'rous swarm
> (Whether attracted by the silence deep
> And pause of rural toil, or sudden struck
> By that instinctive impulse which directs
> More wisely than proud Reason's rule) rush out
> In myriads and take wing : while mingling sounds
> Of distant church-bells and the jangling pan
> Essayed in vain to stop the living cloud."—*Grahame.*

If not attracted, the insects settle on any projection, by preference a bough, and "hanging bee by bee, Make a long sort of rope below the tree" (King), a "pensile cluster," and clinging together, with their queen in the centre, form a ball :—

> " Not closer, orb in orb, conglobed are seen
> The buzzing bees about their dusky queen."—*Pope.*

The "calling" of bees by metallic sounds is often referred to ; thus in Swift, Moore, Young, and Waller :—

S

> " So by the kettle's loud alarm
> The bees are gathered to a swarm."

> " New legions soon
> Pour to the spot like bees of Kamzaroon
> To the shrill timbrel's summons."

> " So swarming bees that on a summer's day,
> In airy rings and wild meanders play,
> Charmed with the brazen sound, their wand'rings end."

> " So wandering bees would perish in the air,
> Did not a sound proportioned to their ear
> Appease their rage, invite them to the hive,
> Unite their force, and teach them how to thrive."

Of the homeward flight of the bee, the load it carries, its satisfied hum, and the directness of its course—a "bee-line" has passed into a proverb for straightness—there are many pretty recognitions. Not that Rogers, casting about for illustrations of the Pleasures of Memory, is specially felicitous : -

> " Who guides the patient pilgrim to her cell?
> Who bids her soul with conscious triumph swell?
> With conscious truth retrace the mazy clue
> Of varied scents, that charmed her as she flew?
> Hail, Memory, hail ! thy universal reign
> Guards the least link of being's glorious chain."

for bees seldom return by the same way that they go.

The mother-bee is never in any sense a sovereign. She is simply an egg-producing machine, which has been constructed by the "horny-handed" toilers of the hive, and (as one may always expect the proletariat to do when they are the masters), they take good care to let their "queen" understand that she is their property and the work of their hands. Every possible care is taken of her as a valuable machine, that she shall not be injured or go astray. She

is stoked with food incessantly, and a certain number of
workers are always in attendance to see to the cleanliness
of every part of their patient egg-layer and her surroundings.
Having made only one perfect female, they are most anxious
to save themselves loss of time and, as they think them, of
earnings, by any accident to her.   But affection there is
none.   When she is getting used up, and either does not
lay eggs fast enough, or begins to lay undersized ones, or
drone-eggs only, they hustle her off the premises, and let
the poor creature die at the door of the hive from cold
and hunger.   Or if by mismanagement there happen to be
superfluous "queens," the horny-handed display an abominable eagerness in having the redundant females done to
death.   They themselves are sexless, or rather are abortive
females, and it is not difficult from a human standpoint to
understand the bitter glee with which these horrid neuters
assist in the murder of the perfect females, who if permitted
to live, would enjoy those privileges of the sex which are
denied to their misformed selves.   And yet, atrocious to
relate, these very neuters are able of their own act, by the
mere fact of having accidentally eaten some of the "royal
jelly," to lay eggs, and these eggs, monstrous ova that they
are, actually produce bees!   Has nature offered anywhere
in all her works a more baffling problem of all-round per-
plexity than this?

Then too the drone, with his big silly head, and his eyes
meeting atop and giving him a foolish blank-eyed look, thinking himself so fine a fellow.   Watch him come out of the hive
on a fine morning.   How cautiously he pokes out his feelers
to see if it be cold or not; how he stops up the doorway
against the industrious neuters, who have been up and at
work since daybreak; how he loiters on the ledge, elaborat-
ing his toilet, exasperating the laden folk who cannot find
room to settle.   And then he thinks of breakfast, and flies
off with a slow heavy droning flight to the nearest flowers.

Among these he spends his days, or goes off to court some "queen" out on her nuptial flight and summoning her suitors. Then the season of courtship and nuptials comes to an end. But the burly drone does not know it, and goes blundering into the hive as usual, and goes to sleep full of honey and fatigued with pleasure. And the horrible awakening! The hive is all astir, and the voices of the workers are hot and hoarse. They are rushing about in all directions, dragging out the drones from where they lie asleep, stabbing them to death, and dragging their bodies on to the sunny ledge, dropping them over on to the ground beneath. The massacre does not take long. For the drones have no stings, and can do no more when the poisoned poniards slip in between their scales than shrill for mercy; which is never granted. So they die in the midsummer of their pleasure.

"She," for in poetry the working neuter is always addressed as feminine, is both the wax-producer, "working her formal rooms in waxen frame" (Spenser), and the honey-gatherer. She—

> " Gives us food that may with nectar vie,
> And wax that does the absent sun supply."—*Waller.* |

Apart from its industry, the bee's skill, both in "distilling" nectar and as a scientific architect, meets with very adequate recognition. Thus :—

> " Spring's unfolded blooms
> Exhaling sweetness, that the skilful bee
> May taste at will, from their selected spoils,
> To work her dulcet food."—*Akenside.*

> " The busy tribe
> Of bees, so emulous, are daily fed
> With heaven's peculiar manna. 'Tis for them,
> Unwearied alchemists, the blooming world
> Nectareous gold distils."—*Hurdis.*

" The bee observe,
She too an artist is, and laughs at man,
Who calls on rules the sightly hexagon
With truth to form, a cunning architect,
Who at the roof begins her golden work,
And builds without foundation.   How she toils,
And still from bud to bud, from flow'r to flow'r,
Travails the livelong day."—*Hurdis.*

Among the legends of flowers there is one that tells how a fair Corinthian was turned in a moment of Olympian pique into a rose, and her crowding suitors into the thorns that defend the beauty of the flowers.   A few of the more forward became insects, and among them was the drone.

But the honey-bee is not one of the number—wherein the ancients showed their wisdom, for bees, being wise insects, do not lurk and hover in roses.   These flowers give an undivided industry to the production of perfumes, and secrete no honey.   The bee wants honey, not perfume. The whole soul of the roses is in attar.   As for the fragrant distillations which help gods and heroes to mead, metheglin and hydromel, they care nothing.   Their laboratories are only for scent.   The parable of this the poets have missed. Yet it is a little curious that such beauty should be honeyless.   The rose is passing fair to the eye, and kind to every breeze that woos, but it has no sweetness at heart : it is a barren loveliness of face and form.   Perhaps as queen of the garden it is becoming that she should not encourage the Ixions of the hive ; and that when the bees, giddy with the perfumed gold dust of the lilies, attempt her charms, she should baffle them with cold, sweet, powderless petals, tempting as Hybla, but "vain as painted fruits."   So the bees, except when on rare occasions pollen fails them, never assail the roses.   But the poets, not unpoetically, give those flowers pre-eminence in the honey-seeker's eyes :—

> "The rose
> That was the queen of all the sunny year,
> She, in whose perfumed halls the wild bee lingers,
> Lightening his toil with song."

The fancy is very beautiful, but in fact the bee, when it happens to go by accident to a rose, does not linger there. On the contrary, it hurries away from the sweet deceiver, grumbling hugely in deep demur.

It is much in the same spirit that the Oriental poets—and Moore, Byron, Shelley, and others after them—represent the rose and nightingale in constant association. It is true that in the East, the bulbuls sing as often in the rose shrubs as in any others; but the poets' association is apart from this; it is proper to fancy that sweets go to the sweet, and the more beautiful to the more beautiful.

So, too, several poets have bees in "jessamine bowers." For myself, I doubt the honey-gatherer's caring for the flower. I know (from personal knowledge) of only one jessamine "bower." Butterflies come there, and sleepy two-winged flies, with large ox-heads and half-transparent bodies. So do the moths with glowing eyes—the wonder of a moth's eyes!—and shoulders wrapped in tippets of fur. But bees, except, I think, in blundering chance, come never. I do not think their tongues can reach the honey at the bottom of the jessamine tube.

But the poets do not extend the error to their other favourite flowers, such as the primrose, violet, or hyacinth, for, except where one talks of the bee, "o'er sun-flowers singing" (by which the heliotrope, it may be, is intended), the flowers of the garden are not associated with them.

Next only to the song of birds, the voice of the bee attracts the poets. Nor indeed could it well be otherwise, for who could write of flowers, or of sunshine, or of summer, without mention of it, and mentioning it without admiration? The great majority of the poets accept its "drowsiness" as the chief feature of the bee's voice. It is "deep music, pleasant melody to the weary traveller" (Southey), "lulling" (Leyden), a "lullaby" (Charlotte Smith), "inviting sleep sincere" (Thomson), "low vexed song" (W. Morris), "sad murmuring tune" (W. Morris), "plaintive, inviting to rest," "sleep-inviting," "drowsy" (Shelley). But Eliza Cook goes too far when she says that bees "go droning by, and hum themselves to sleep," which is exactly what they do not do. Far more accurate seems Shelley's "soft strain," and Milton's "singing at her work." The bee is not itself drowsy, though its hum decidedly "invites to sleep," suggestive also above, all other sounds, of peace. Rural poets, Clare, Grahame, Hurdis, Bloomfield, and no doubt others, specially connect it with the Sabbath day, more struck perhaps by it on that day "from the pause of rural toil."

It is in many poets a note of gladness, a "joyous chorus," and eloquent of gratitude :—

> " And lo ! within my lonely bow'r
> Th' industrious bee from many a flow'r
> Collects the balmy dews.
> For me, she says, the gems are born,
> For me their silken robe adorn,
> Their fragrant breath diffuse."

> " Sweet murmurer, may no rude storm
> This hospitable scene deform,
> Nor check thy gladsome toils :
> Still may the buds unsully'd spring.
> Still show'rs and sunshine court thy wing
> To these ambrosial spoils."—*Akenside.*

> " On hanging cobwebs shone the dew,
> And thick the wayside clovers grew,
> The feeding bee had much to do,
> So fast did honey-drops exude ;
> She sucked and murmured, and was gone,
> And lit on other blooms anon,
> The while I learned a lesson on
> The source and sense of gratitude."—*Jean Ingelow.*

> " 'The vagrant bee that sings
>   For what he gets thereby,
>   Nor comes unless he brings
>   His pocket on his thigh."—*Mary Howitt.*

and Shelley, addressing Night, has the line, " My sweet child Sleep, the filmy-eyed, Murmured like a noontide bee."

Hurdis sums up from nature the various tones of the different bee voices :—

> " Where the cherry spreads its flowery tufts,
>   'Tis pleasure to survey the snowy pomp,
>   And pause in contemplation of the hum
>   Of mingled bees industrious, that invade
>   And rifle in succession every flower ;
>   Some large and gifted with the voice profound
>   Of mellow bass, some with the loftier pipe
>   Of tenor soft, of small soprano some,
>   That fancy oft may deem she hears distinct,
>   The sweet coincidence of fellow-tones
>   Producing harmony's full chord divine ; "

One or two eccentricities of descriptions are worth noting. " Bagpiping " is Montgomery's epithet for the bee's hum, while Mackay at least half-a-dozen times calls the bee "trumpet-toned," its voice that of a " booming trumpet," and speaks of the " buzzing of bees in trumpet-tones," and, most ridiculous horror of all (though " buzzing " in a " trumpet-tone " is hard to beat)—

> " Sounding a trump like a martial call
>   On a clarion of brass."

Could anything be more odious as a description of the sound that Keats calls " a bee's demur," and which makes the foxgloves musical ? The trumpet idea had been prettily used by Mrs. Hemans as a " drowsy bugle," and Hood, before her, speaks of the fairies and

> " Their horns of honeysuckle blooms
>   Sounding upon the air most soothing soft,
>   Like humming bees."

while, earlier still, Faber has the exquisite line in which
he hears the bees

> " Blowing the trumpets of the eglantine."

That he meant woodbine does not matter ; for on this point
he errs with Milton, and the fancy is beautiful enough
to justify many errors of petty fact. But "the trumpets
of the eglantine" are a very different matter from "a
trump like a clarion of brass." Both Byron and Keats
rank the bee's voice among the most beautiful of sounds,
with "the voice of girls, the song of birds, The lisp of
children and their earliest words."

In Moore's fantastic botany, such strange blossoms as
the Arura "just oped," saffron (better known as the crocus)
in full flower, the Nilia,

> " Lulling is the song
> Of Indian bees at sunset, when they throng
> Around the fragrant Nilia, and deep
> In its fragrant blossoms hum themselves to sleep ; "

and the Sihu-thorn are met with tempting the hive-people.
If the poet had really known the East, with its legends
of flower and bee, or even the mango only, when the
love-god Camdeo takes up his bow in the Spring, and goes
forth, his arrows barbed with mango-blossom, his bow-string
a cord of bees,

> " He bends the luscious cane, and twists the string
> With bees, how sweet ! but, ah ! how keen their sting."
> —*Sir W. Jones.*

It is only in Moore, too, that we find the "voluptuous "
and the "sated" bee asleep in "the languished noon "
in the hearts of flowers—a natural license of imagination,
and, curiously enough, true to nature, as bees are often
found drugged to sleep in the bosom of the flowers they
went to rifle.

The "bee's sweet alchemy" is a sufficiently engaging

phrase, and no doubt the assiduous insect, seeking the "mellifluous streams, its balmy spoils," and "pleased with the sweet variety of many-tempered flowers," is a wonder-worker, for it blends the good with the bad into the gracious "nectar of the hive."

For this the poets perpetually admire it :—

> " Hush'd at her voice, pert Folly's self is still,
> And dulness wonders while she drops her quill
> Like the armed bee, with art most subtly true,
> From pois'nous vice she draws a healing dew."—*Pope.*

> " The labouring bees,
> That in the Summer heap their winter's food,
> Plied to their hives sweet honey from those flowers,
> Whereout the serpent strengthens all his powers."—*Greene.*

> " Nature, that gives the bee so feat a grace
> To find honey of so wondrous fashion,
> Hath taught the spider out of the same place
> To fetch poison by strange alteration."—*Wyatt.*

> " E'en bees, the little alms-men of Spring flowers,
> Know there is richest juice in poison-flowers."—*Keats.*

Yet the legend is as old as language, and as widespread as sunlight, that bees can make poison as well as honey; and sober history records instances where men, and even armies, have suffered from eating rashly from combs stored with dangerous juices.

> " Even as those bees of Trebizond,
> Which from the sunniest flowers that glad
> With their pure smiles the gardens round,
> Draw venom first that drives men mad."—*Moore.*

Once when I was at Midhurst, I tasted a bottle of mead which my host had found in his cellar, overlooked for many years. It was my only experience of "bees' nectar," and it more than satisfied all the expectations that romance had encouraged. As soon as the cork was withdrawn—it was a heavy August day, with the windows wide open to the clover-field on one side and the pine-woods on the other— there floated through the house, and hung about it after- wards for hours, a most wonderful fragrance of herbs and flowers.

A year later I was in the Mauritius, walking "sous les filaos" in the gardens at Pampelmousse (is that the way to spell it?), when a "something" of the same scent reached me; and once again, in the offices of the Grande Chartreuse, in Carey Street, off the Strand, when "the monks" were showing me their museum, I smelt it, and said "Mead."

"What is it like?" Certainly not the drink that Phillips, in his Cider, thus odiously describes as a gargle :—

> " The Britons squeeze the work
>   Of sed'lous bees, and mixing odorous herbs,
>   Prepare balsamic cups, to wheezing lungs
>   Medicinal, and short-breath'd ancient sires."

But what it is like I cannot say :—

> " The bee shall sip the fragrant dew from flowers,
>   To give metheglin for his morning hours ; "

but no bank of wild flowers ever yet gave me the odours of metheglin. The Vikings knew it well; so in Gray's poem on the descent of Odin :—

> " Mantling in the goblet, see
>   The pure bev'rage of the bee.
>   O'er it hangs the shield of gold,
>   'Tis the drink of Balder bold."

And again, in his Death of Noël, we have it on the poet's word how—

> " From the golden cup they drink
>   Nectar that the bees produce."

But how do the bees, "collecting the various odours

of the Spring," get that wondrous heather and pine, that hyacinth and underbloom, that heliotrope and thyme, all together into one comb?

> "As bees mixed nectar draw from fragrant flowers "

is trite enough, even for Young; but in mead the nectar is *not* mixed.

Each flower is there authentic and indisputable. Waller is utterly wrong who sings—

> " Sweetest, you know the sweetest of things
>   Of various flowers the bees do compose,
>   Yet no particular taste it brings
>   Of violet, woodbine, pink, or rose."

For call it what you will, metheglin or hydromel, soma or mead, the compound is "nectar, drink of gods," the whole woodside in a bottle, Pan with all his poems in a glass.

You may think I rhapsodise. It is not so, as others who have drunk mead twenty years old will tell you.

Southey addresses the bee as a miser, warning it that it will not live to enjoy the fruits of its toil :—

> " Thou art a miser, thou busy busy bee,
>   Late and early at employ ;
> Still on thy golden stores intent,
>   Thy summer in reaping and hoarding is spent
>   What thy winter will never enjoy :
> What lesson this for me, thou busy busy bee !
>
> Little dost thou think, thou busy busy bee,
>   What is the end of thy toil,
> When the latest flowers of the ivy are gone,
>   And all thy work for the year is done,
>   Thy master comes for the spoil :
> Woe then for thee, thou busy busy bee !"

The fact is true enough, but not always for the reason which Southey gives. For it is a deplorable fact that a worker-bee lives only six weeks, so that all the earlier toilers are dead before the time arrives for consuming the stored honey.   Against the cruelty of suffocating bees to get at the combs (a practice now fast going out of fashion), Thomson, in his Autumn, protests :—

> " Ah, see where, robb'd and murder'd in that pit,
>   Lies the still-heaving hive, at evening snatch'd
>   Beneath the cloud of guilt concealing night,
>   And fix'd o'er sulphur ; while not dreaming ill,
>   The happy people in their waxen cells
>   Sat tending public cares, and planning schemes
>   Of temperance, for winter poor, rejoiced
>   To mark, full flowing round, their copious store.
>   Sudden the dark oppressive steam ascends,
>   And, used to milder scent, the tender race
>   By thousands tumble from their honeyed domes,
>   Convolv'd and agonising in the dust.
>   And was it then for this you roam'd the Spring
>   Intent from flower to flower ; for this you toil'd
>   Ceaseless the burning Summer heats away ;
>   For this in Autumn search'd the blooming waste,
>   Nor lost one sunny gleam?   For this sad fate?
>   Ah, man, tyrannic lord, how long, how long
>   Shall prostrate nature groan beneath your rage,
>   Awaiting renovation, when obliged
>   Must you destroy? Of their ambrosial food
>   Can you not borrow, and in just return
>   Afford them shelter from the wintry winds ;
>   Or, as the sharp year pinches, with their own
>   Again regale them on some smiling day?
>   See where the stony bottom of their town,
>   Looks desolate and wild ; with here and there
>   A helpless member, who the ruin'd state
>   Survives, lamenting, weak, cast out to death."

while Grahame pleads for their being fed in the early Spring :—

> " Now profit prompts, if pity ask in vain,
>   To save the falling state ; nor large the boon
>   They crave—the refuse of the summer spoil,
>   Or syrup of the cane in bour-tree[1] trough
>   Pushed softly in, will help them, till the down
>   Hang on the willow tree, than which no flower
>   Yields fruit more grateful to the frugal tribe."

---

[1] Elder.

These wretched workers are no sooner out of their cells than they go to work—at first in the hive, on menial jobs, "doing the chores," and then in the toil of collecting honey, pollen, and propolis. Their bodies therefore are mere portmanteaus. Inside they are bags for holding fluid honey, and outside they have six pockets, into which they exude wax, and two trough-like receptacles on their thighs for holding pollen. Miserable little omnibuses. All day long they fly from "the Bank" to the flower, and from the flowers back to "the Bank," full inside and full out. Even the fur upon their backs is used to augment their loads, for the pollen of the flowers catches in it and sticks ; and when they get home they are tooth-combed out by attendants. It is a dreadful life, and a very immoral one, for what is the end of it? A comfortable winter of leisure and luxury, in a warm hive, with plenty of companions and abundant food? Not a bit of it. The toilers who secrete the wax and build the cells, who collect the honey and the bee-bread to fill those cells "against winter," are all dead before winter comes—

> " So bees to feed another's need,
> From flowers doe honey gather."

They live only about six weeks. And if by any accident they should come to harm, be maimed, or sick, they are either murdered or thrust out from the hive, to die as they please on the ledge outside the door.

Sic vos non vobis mellificatis apes. Yet, sexless, soulless, as these assiduous little "neuters" are, there is a great pathos in the thought that nothing but the feeding which they might have had separates them from the life of a "mother-bee," who lives several years, is always attended to with unceasing care, and defended from every vexation and want, and who, by the accident of a different diet for a week, is endowed with every feminine susceptibility, and achieves the highest ambitions of her sex—is repeatedly married, numerously widowed, and multitudinously a mother. And all this time—the supreme bliss of it ! is the one and only "woman" in the whole hive, the object of every anxiety, the centre of all attention, the cherished "luck" of the commonwealth and its palladium. Poor little neuters. No wonder with your half-awakened female instincts you rush so madly backwards and forwards about your work—is it to keep thought away?—and so soon wear out the great true hearts beating in your misformed bodies.

As telling the hours of the day, even the days of the week, and the months of the year, the poets have constructed a bee-calendar—

> " The light of morn, with hum of bees,
> Steals through the verdurous matting of fresh trees."—*Keats.*

And the poets meet it busy among the heather, as they stand upon the hill to greet the rising sun (Hurdis) :—

> " The active bee out at early morn,
> Just as the opening flowers are born."—*Charlotte Smith.*

It is Noon and—

> " Glossy bees do fieldward pass."—*Keats.*

and—

> " Beneath the fervour of the noontide beam
> All nature's works in placid stillness pause,
> Save man and his joint-labourer,—the horse,
> The bee, and all the idly busy insect tribe."—*Grahame.*

> " It was noon, and in flowers that languished around,
> In silence reposed the voluptuous bee."

It is Evening :—

> " The sun is low sinking behind the trees,
> And crossing the path hum home the bees."—*Clare.*

It is Sunday—no day of rest for the little drudge—and the bee's hum sounds louder than usual, marking the Sabbath with more audible canticles. So in Mary Howitt's

"Sunday," when all else around bespeaks the seventh day, the bee "hums in the clover," as if,—from the contrasting quiet of the fields and lanes, the absent cries of men at work with their horses, of the shepherd and his dogs, the creaking of deep-cutting wheels,—the troubadour of the flowers were in fuller voice.   So too in Grahame's "Sabbath."

It is Winter and—

> " The bees in hives idly wait,
> The call of early Spring."—*Cowper.*

A warm day in February tempts them forth :—

> " Sometimes, deceived by promise premature
> Of spring's approach, or pinched by empty combs,
> Forth from the hive some straggling bees will peep,
> And buzzing on the outside of their porch
> Will try their wings, but not attempt to fly."—*Grahame.*

But Spring really comes :—

> " Winter is past—the little bee resumes
> Her share of sun and shade, and o'er the lea
> Hums her first hymnings to the flowers' perfumes,
> And wakes a sense of gratefulness in me."—*Clare.*

> " The empty bee, that lately bore
> Into the common measure all her store,
> Flies 'bout the painted fields with nimble wing,
> Deflow'ring the fresh virgins of the Spring."—*Carew.*

And so the Summer, " with its flowers and bees," when the swarms "rush out in myriads and take wing," and moorland and meadow and blossoming tree are musical with "the murmuring of innumerable bees."

The "wild" bee of poetry is the humble-bee.   Sometimes it is specified by name, as in Clare's " drone of heavy humble-bees," and Cowper's "drone-pipe of heavy humble bees," and in Carew's voice of spring that "wakes in hollow tree the drowsie Cuckow and the Humble-bee."

Elsewhere it is only "the wild-bee," but there can be no doubt of the insect intended—"the bees that in the neighbouring trunk make honey-hoards" (Coleridge), the "wild bee voyaging deep-toned" (Grahame), whose "deep drowsy murmurs pass, like a low thrill of harp-strings, through the grass" (Mrs. Hemans), "with sudden buzz slung past the ear" (Leyden):—

> " the buzz
> Angrily-shrill of moss-entangled bee,
> That soon as loosed booms with full twang away."—*Hurdis.*

There are in England many, more than a dozen, kinds of "humble-bees," and two of these are mentioned by name in verse. In "Midsummer Night's Dream" Bottom asks Cobweb to "kill him a red-tipped humble-bee on the top of a thistle : and good monsieur, bring me the honey-bag," and in "Alls Well that Ends Well" it is the "red-tailed humble-bee." This beautiful and familiar insect is the "lapidary" bee, which builds by preference among or between stones. It is an excitable insect, and fierce in defence of its home, and is probably the bee referred to by Grahame.

> " Should they find a sable swarm's retreat,
> Deep-earthed, the mining spade must lay it bare ;
> Nor unresisting do the inmates yield,
> Their little state : forth at the first alarm
> They swarming rush, and chacing in long train
> The flying foe, deal sharp not deadly wounds ;
> Rallied at length, the assailants to the charge,
> With doublets doffed attack the stinging tribes."

Another is called the "yellow-banded and "yellow-girted," and is the equally common "ground" humble-bee. Poets have another variety, "the brown-bee," but I cannot identify it.

> " The mower
> Draws from the grass the brown-bee's honied nest."—*Leyden.*

> " The younger wights with busy eye explore
> The foggage, where, concealed with mickle art,
> The brown-bee's cups in rude formed clusters lie."—*Grahame.*

T

William Morris too has "o'er the bindweed the brown bee passes." Or is this only another form of the "brown bird," so frequent in his verse?

There is also the "mountain-bee." For instance :—

> "When hums the mountain-bee in May's glad ear."—*Wordsworth.*

> "The mountain-bee was roving with his minstrelsy across
> The scented wild weeds."—*Campbell.*

> "Wild as a mountain-bee."—*Hurdis.*

Can this be a poetic "variety" only, like the "mountain-lark" (possibly an error in sound for "mounting lark," which is so very common a phrase in verse), or "mountain-wolf," as superadding a special "wildness" to the species?

The "poetical" character of the bee is summed up excellently in Churchill's "Gotham" :—

> "The hive is up in arms, expert to teach,
> Nor proudly to be taught unwilling; each
> Seems from her fellow a new zeal to catch;
> Strength in her limbs, and on her wings despatch,
> The bee goes forth; from herb to herb she flies,
> From flow'r to flow'r, and loads the lab'ring thighs,
> With treasured sweets, robbing those flow'rs which, left,
> Find not themselves made poorer by the theft.
> Ne'er doth she flit on pleasure's silken wing,
> Ne'er doth she loit'ring let the bloom of spring
> Unrifled pass, and on the downy breast
> Of some fair flower indulge untimely rest.
> Ne'er doth she, drinking deep of those rich dews
> Which chymist night prepared, that faith abuse
> Due to the hive, and selfish in her toils,
> To her own private use convert the spoils.
> Love of the stock first called her forth to roam,
> And to the stock she brings her booty home."

This loyalty finds frequent recognition. Thus in Crabbe—

> "Like wounded bees that at their home arrive,
> Slowly and weak but labouring for the hive."

So in Gay's fable of "the degenerate bees," when certain

members of the hive try to corrupt the community by describing the pleasures of idleness and luxury, "a stubborn bee" gets up, and "with honest indignation warm" replies to their base counsels :—

> " Shall luxury corrupt the hive
> And none against the torrent strive?
> Exert the honour of your race ;
> He builds his rise on your disgrace.
> 'Tis industry our state maintains ;
> 'Twas honest toil and honest gains
> That raised our sires to power and fame.
> Be virtuous ; save yourselves from shame.
> Know that in selfish ends pursuing,
> You scramble for the public ruin."

" Vagrant " and " vagabond " are, next to " busy " and its synonyms, the most frequent epithets of the honey-seeking bee, a libertine for the purpose chartered. It is a lover with a roving commission, and as a trifler with the affections of flowers, far more seriously detrimental than the butterfly, the insect which is so perpetually being reproached by poets for fickle love and wandering fancy. The simile of inconstancy is obviously on the surface, and at once suggested by the habits of the bee; but it is very curious that this practical, prosaic, honey-grubbing little brown fly should live in verse as a far more dangerous flirt than that gallant brilliant courtier the butterfly, and be held up as a warning to pretty blossoms of the dangers to be apprehended from "deflowering " insects.   As thus in Moore :—

> " *He.* What the bee is to the flowret
>   When he looks for honey dew,
>   Through the leaves that close embower it,
>   That, my love, I'll be to you.
>
> " *She.* What the bank with verdure glowing
>   Is to waves that wander near,
>   Whispering kisses, while they're going,
>   That I'll be to you, my dear.

"*She.* But they say the bee's a rover,
Who will fly when sweets are gone,
And, when once the kiss is over,
Faithless brooks will wander on.

"*He.* Nay, if flowers will lose their looks,
If sunny bank will wear away,
'Tis but right that bees and brooks
Should sip and kiss them while they may."—*Moore.*

" Love the result is of all the graces
Which flow from a thousand several faces."—*Waller.*

" With the bee he doth rejoice
Every minute to change choice,
Counting he were then in bliss,
If that each fair face were his."—*Greene.*

Yet here and again, with an effort of fancy, this destroying angel of the peace of flowers, the breaker of the hearts of lilies, is spoken of as constant in love :—

" Thus, as the bee from bank to bower
Assiduous sips at every flower,
But rests on none till that be found
Where most nectareous sweets abound."—*Cowper.*

" The bee through many a garden roves,
And hums his lay of courtship o'er ;
But when he finds the flower he loves,
He settles there and hums no more."—*Moore.*

So it will be seen that in character the bee has violent contrasts, now with honest music assiduous in toil, now frivolously trifling with the innocence of virgin flowers. As busy, it is held up to the idle grasshopper as an example, otherwise it is lectured as listless, drowsy, vagrant.

Allan Ramsay describes the bee as a prig of the first magnitude :—

" Before her hive a paughty Bee
Observ'd a humble midding Flie,

> And proudly speer'd what brought her there,
> And with what front she durst repair
> Among the regents of the air.
> ' It sets ye well,' the Flie reply'd,
> ' To quarrel with sic saucy pride !
> They're daft indeed who've ought to do
> With thrawn contentious fowk like you.'
> ' Why, scoundrel, you ! ' return'd the Bee,
> ' What nation is sae wise as we ?
> Best laws and policy is ours,
> And our repast the fragrant flowers.
> No sordid nasty trade we drive,
> But with sweet honey fill the hive—
> Honey maist gratefu' to the taste,
> On which the gods themselves may feast.
> Out of my sight, vile wretch ! whose tongue
> Is daily slacking throw the dung ;
> Vile spirits, filthily content
> To feed on stinking excrement ! '
> The Flie replied in sober way,
> ' Faith, we maun live as well's we may.
> Glad poverty was ne'er a vice,
> But sure ill-natur'd passion is.
> Your honey's sweet, but then how tart
> And bitter's your malicious heart.
> In making laws you copy heaven,
> But in your conduct how uneven !
> To fash at ony time a fae,
> Ye'll never stick yer'sells to slae,
> And skaith yer'sell mair sickerly
> Than e'er ye can your enemy.
> At that rate ane had better have
> Less talents, if they can behave
> Discreet, and less their passions' slave.' "

The bee has no large place in classic tale or fable.

> " In such a palace Aristæus found
> Cyrene, when he bore the plaintive tale
> Of his lost bees to her maternal ear."

Cowper refers here to a legend that has many versions, and though it is difficult and not a little impious to try nowadays to decide which is right, the main facts appear to be these. Aristæus was a young prince who kept bees. He fell in love with Eurydice, wife of Orpheus, and one

day, in the prosecution of his suit, that is to say, when
chasing her across country, as appears to have been the
fashion of love-making in those days, the lady was so
stung by his bees that she died. Orpheus thereupon
appealed to the gods, who in response swept all the bees
off the face of Greece. Aristæus complained to his mother,
who in turn appealed to her father, Proteus, who showed
the youth how to get back his beloved insects. He
sacrificed a number of fat kine to the ghost of Eurydice, and
from their carcases issued swarms of bees.

Hence Butler's lines :—

> " The learned write, an insect ' breeze '
> Is but a mongrel sort of bees,
> That fall before a storm on cows,
> And sting the founders of their house,
> From whose corrupted flesh that breed
> Of vermin did at first proceed."

That bees ballasted themselves with pebbles and other
make-weights when flying against the wind is an artifice
of which they share the credit with cranes. "Bees that
are employed in carrying off honey choose always to have
the wind with them if they can. If haply there do arise
a storm while they be abroad, they catch up some little
stony grit, to balance and poise themselves against the
wind. (Some say that they take it and lay it upon their
shoulders). And withal they fly low by the ground under
the wind when it is against them, and keep along the
bushes, to break the force thereof"—all of which, except
the line in brackets, is an excellent description enough of
the flight of some of the carnivorous bees as they come
home laden with the numbed honey-bees, spiders, or caterpillars, with which they store their egg-cells, as food for
their grubs when hatched.

Once upon a time there was in France an Order of
the Mouche à Miel, and the oath taken by knights when

joining was sufficiently curious. "I swear, by the bees of Hymettus, to be loyal and obedient to the Directress of the Order, to wear all my life the medal of the Fly, and to fulfil, as long as I live, the statutes of the Order ; and, if I am false to my oath, may honey turn in my mouth to gall, wax in my hands to grease, the flowers I gather to nettles, and may wasps and hornets be for ever stinging me."

In metaphor, simile, and moral, the poetic bee abounds. Thus swarming suggests colonisation :—

> "Of Britain to invite her to cast off
> Her swarms, and in succession send them forth."

> " Their surcease grew so great, as forced them at last
> To seeke another soyle (as Bees doe when they caste)."—*Drayton.*

Young on the same simile introduces the key-and-saucepan :—

> " Each soul
> That ever animated human clay
> Now wakes, is on the wing ; and when, oh ! where
> Will the swarm settle ?—When the trumpet's call
> As sounding brass collects us, round Heaven's throne
> Conglobed we bask in everlasting day
> (Paternal splendour) and adhere for ever."

Its voice affords similes for populous haunts of man :—

> " And there the murmur of the neighbouring quay
> Dwells like the humming of a drowsy bee."—*Faber.*

> " Bee-like, bubbling, busy hum
> Of cities."—*Byron.*

Sweet soft speech, mellifluous :—

> "The working bees' soft melting gold,
> That which the waxen mines enfold,
> Flows not so sweet as do the tones
> Of his [1] tuned accents."—*Crashaw.*

---

[1] Cupid's.

> " On his eloquent accents fed and hung
>   Like bees on mountain flowers."—*Shelley.*

> " Her words were like a stream of honey fleeting,
>   The which doth softly trickle from the hive."—*Spenser.*

"Plus mellis quam fellis" was the appropriate motto of the bee-crest of the eloquent St. Ambrose.

INDEX.

THE END.

PRINTED BY BALLANTYNE, HANSON AND CO.
EDINBURGH AND LONDON.

Telegraphic Address—BOOKSTORE, LONDON.   Telephone No.—8534 CENTRAL.

AN ALPHABETICAL CATALOGUE
OF BOOKS IN FICTION AND
GENERAL LITERATURE
PUBLISHED BY
CHATTO & WINDUS
III ST. MARTIN'S LANE
CHARING CROSS
LONDON, W.C.
[SEPT., 1903.]

A B C (The) of Cricket: a Black View of the Game. (26 Illustrations.) By HUGH FIELDING. Demy 8vo, 1s.

Adams (W. Davenport), Works by.
A Dictionary of the Drama: being a comprehensive Guide to the Plays, Playwrights, Players, and Playhouses of the United Kingdom and America, from the Earliest Times to the Present Day. Crown 8vo, half-bound, 12s. 6d. *[Preparing.*
Quips and Quiddities. Selected by W. DAVENPORT ADAMS. Post 8vo, cloth limp, 2s. 6d.

Agony Column (The) of 'The Times,' from 1800 to 1870. Edited, with an Introduction, by ALICE CLAY. Post 8vo, cloth limp, 2s. 6d.

Alden (W. L.).—Drewitt's Dream. Crown 8vo, cloth, gilt top, 6s.

Alexander (Mrs.), Novels by. Post 8vo, illustrated boards, 2s. each
Maid, Wife, or Widow? | Blind Fate.
Crown 8vo, cloth, 3s. 6d. each; post 8vo, picture boards, 2s. each.
Valerie's Fate. | A Life Interest. | Mona's Choice. | By Woman's Wit.
Crown 8vo, cloth 3s. 6d. each.
The Coat of her Pride. | Barbara, Lady's Maid and Peeress. | A Fight with Fate.
A Golden Autumn. | Mrs. Crichton's Creditor, | The Step-mother.
| A Missing Hero.

Allen (F. M.).—Green as Grass. Crown 8vo, cloth, 3s. 6d.

Allen (Grant), Works by. Crown 8vo, cloth, 6s. each.
The Evolutionist at Large. | Moorland Idylls.
Post-Prandial Philosophy. Crown 8vo, art linen, 3s. 6d.
Crown 8vo, cloth extra, 3s. 6d. each; post 8vo, illustrated boards, 2s. each.
Babylon. 12 Illustrations. | The Devil's Die. | The Duchess of Powysland.
Strange Stories. | This Mortal Coil. | Blood Royal.
The Beckoning Hand. | The Tents of Shem. | Ivan Greet's Masterpiece.
For Maimie's Sake. | The Great Taboo. | The Scallywag. 24 Illusts.
Philistia. | Dumaresq's Daughter. | At Market Value.
In all Shades. | Under Sealed Orders. |
The Tents of Shem. POPULAR EDITION, medium 8vo, 6d.

Anderson (Mary).—Othello's Occupation. Crown 8vo, cloth, 3s. 6d.

Antrobus (C. L.), Novels by. Crown 8vo, cloth, gilt top, 6s. each.
Quality Corner. | Wildersmoor. | The Wine of Finvarra.

Appleton (G. Webb).—Rash Conclusions. Crown 8vo, cloth, 3s. 6d.

Arnold (Edwin Lester), Stories by.
The Wonderful Adventures of Phra the Phœnician. Crown 8vo, cloth extra, with 12 Illustrations by H. M. PAGET, 3s. 6d.; post 8vo, illustrated boards, 2s.
The Constable of St. Nicholas. With Frontispiece by S. L. WOOD. Crown 8vo, cloth, 3s. 6d.; picture cloth, flat back, 2s.

Art (The) of Amusing: A Collection of Graceful Arts, Games, Tricks, Puzzles, and Charades. By FRANK BELLEW. With 300 Illustrations. Crown 8vo, cloth extra, 4s. 6d.

Artemus Ward's Works. With Portrait and Facsimile. Crown 8vo, cloth extra, 3s. 6d.—Also a POPULAR EDITION post 8vo, picture boards, 2s.

Ashton (John), Works by. Crown 8vo, cloth extra, 7s. 6d. each.
Humour, Wit, and Satire of the Seventeenth Century. With 82 Illustrations.
English Caricature and Satire on Napoleon the First. With 115 Illustrations.
Social Life in the Reign of Queen Anne. With 85 Illustrations. Crown 8vo, cloth, 3s. 6d.
Crown 8vo, cloth, gilt top, 6s. each.
Social Life under the Regency. With 90 Illustrations.
Florizel's Folly: The Story of GEORGE IV. With Photogravure Frontispiece and 12 Illustrations.

Bacteria, Yeast Fungi, and Allied Species, A Synopsis of. By
W. B. GROVE, B.A. With 87 Illustrations. Crown 8vo, cloth extra, 3s. 6d.

Bardsley (Rev. C. Wareing, M.A.), Works by.
English Surnames: Their Sources and Significations. Crown 8vo, cloth, 7s. 6d.
Curiosities of Puritan Nomenclature. Crown 8vo, cloth, 3s. 6d.

Barr (Robert), Stories by. Crown 8vo, cloth, 3s. 6d. each.
In a Steamer Chair. With Frontispiece and Vignette by DEMAIN HAMMOND.
From Whose Bourne, &c. With 47 Illustrations by HAL HURST and others.
Revenge! With 12 Illustrations by LANCELOT SPEED and others.
A Woman Intervenes. With 8 Illustrations by HAL HURST.

Crown 8vo, cloth, gilt top, 6s. each.
The Unchanging East: Notes on a Visit to the Farther Edge of the Mediterranean.
A Prince of Good Fellows. With 15 Illustrations by EDMUND J. SULLIVAN.

Barrett (Frank), Novels by.
Post 8vo, illustrated boards, 2s. each; cloth, 2s. 6d. each.

The Sin of Olga Zassoulich.	**John Ford;** and His Helpmate.
Between Life and Death.	**A Recoiling Vengeance.**
Folly Morrison. | **Little Lady Linton.**	**Lieut. Barnabas.** | **Found Guilty.**
A Prodigal's Progress. |**Honest Davie.**	**For Love and Honour.**

Crown 8vo, cloth, 3s. 6d. each; post 8vo, picture boards, 2s. each; cloth limp, 2s. 6d. each.
Fettered for Life. \| **The Woman of the Iron Bracelets.** \| **The Harding Scandal**
A Missing Witness. With 8 Illustrations by W. H. MARGETSON.

Crown 8vo, cloth, 3s. 6d. each.
Under a Strange Mask. With 19 Illusts. by E. F. BREWTNALL. \| **Was She Justified?**
Lady Judas. Crown 8vo, cloth, gilt top, 6s.

Besant (Sir Walter) and James Rice, Novels by.
Crown 8vo, cloth extra, 3s. 6d. each; post 8vo, illustrated boards, 2s. each; cloth limp, 2s. 6d. each.

Ready-Money Mortiboy.	**This Son of Vulcan.**	**The Seamy Side.**
The Golden Butterfly.	**The Monks of Thelema.**	**The Case of Mr. Lucraft.**
My Little Girl.	**By Celia's Arbour.**	**'Twas in Trafalgar's Bay.**
With Harp and Crown.	**The Chaplain of the Fleet.**	**The Ten Years' Tenant.**

. There are also LIBRARY EDITIONS of all excepting the first two. Large crown 8vo, cloth, 6s. each.

Besant (Sir Walter), Novels by.
Crown 8vo, cloth extra, 3s. 6d. each ; post 8vo, illustrated boards, 2s. each : cloth limp, 2s. 6d. each.
All Sorts and Conditions of Men. With 12 Illustrations by FRED. BARNARD.—Also the
LARGE TYPE, FINE PAPER EDITION, pott 8vo, cloth, gilt top, 2s. net ; leather, gilt edges, 3s. net.
The Captains' Room, &c. With Frontispiece by E. J. WHEELER.
All in a Garden Fair. With 6 Illustrations by HARRY FURNISS.
Dorothy Forster. With Frontispiece by CHARLES GREEN.
Uncle Jack, and other Stories. \| **Children of Gibeon.**
The World Went Very Well Then. With 12 Illustrations by A. FORESTIER.
Herr Paulus: His Rise, his Greatness, and his Fall. \| **The Bell of St. Paul's.**
For Faith and Freedom. With Illustrations by A. FORESTIER and F. WADDY.
To Call Her Mine, &c. With 9 Illustrations by A. FORESTIER.
The Holy Rose, &c. With Frontispiece by F. BARNARD.
Armorel of Lyonesse: A Romance of To-day. With 12 Illustrations by F. BARNARD.
St. Katherine's by the Tower. With 12 Illusts by C. GREEN.—Also in picture cloth, flat back, 2s.
Verbena Camellia Stephanotis, &c. With a Frontispiece by GORDON BROWNE.
The Ivory Gate. \| **The Rebel Queen.**
Beyond the Dreams of Avarice. With 12 Illustrations by W. H. HYDE.
In Deacon's Orders, &c. With Frontispiece by A. FORESTIER. \| **The Revolt of Man.**
The Master Craftsman. \| **The City of Refuge.**

Crown 8vo, cloth, 3s. 6d. each.
A Fountain Sealed. \| **The Changeling.** \| **The Fourth Generation.**

Crown 8vo, cloth, gilt top, 6s. each.
The Orange Girl. With 8 Illustrations by F. PEGRAM.
The Lady of Lynn. With 12 Illustrations by G. DEMAIN-HAMMOND.
No Other Way. With 12 Illustrations by CHARLES D. WARD.

POPULAR EDITIONS, medium 8vo, 6d. each.

All Sorts and Conditions of Men.	**The Chaplain of the Fleet.**
The Golden Butterfly.	**The Orange Girl.**
Ready-Money Mortiboy.	**Children of Gibeon.**

The Charm, and other Drawing-room Plays. By Sir WALTER BESANT and WALTER H. POLLOCK.
With 50 Illustrations by CHRIS HAMMOND and JULE GOODMAN. Crown 8vo, cloth, 3s. 6d.

Fifty Years Ago. With 144 Illustrations. Crown 8vo, cloth, 3s. 6d.
The Eulogy of Richard Jefferies. With Portrait. Crown 8vo, cloth, 6s.
Sir Richard Whittington. With Frontispiece. Crown 8vo, art linen, 3s. 6d.
Gaspard de Coligny. With a Portrait. Crown 8vo, art linen, 3s. 6d.
The Art of Fiction. Fcap. 8vo, cloth, red top, 1s. net.
As We Are and As We May Be. Crown 8vo, buckram, gilt top, 6s.
Essays and Historiettes. Crown 8vo, buckram, gilt top, 6s.

Demy 8vo, cloth, 7s. 6d. each.
London. With 125 Illustrations.
Westminster. With an Etched Frontispiece by F. S. WALKER, R.E., and 130 Illustrations by
WILLIAM PATTEN and others.
South London. With an Etched Frontispiece by F. S. WALKER, R.E., and 118 Illustrations.
East London. With an Etched Frontispiece by F. S. WALKER, and 55 Illustrations by PHIL
MAY, L. RAVEN HILL, and JOSEPH PENNELL.
Jerusalem : The City of Herod and Saladin. By WALTER BESANT and E. H. PALMER. Fourth
Edition. With a new Chapter, a Map, and 11 Illustrations.

Baring Gould (Sabine, Author of 'John Herring,' &c.), Novels by.
Crown 8vo, cloth extra, 3s. 6d. each; post 8vo, illustrated boards, 2s. each.
Red Spider. | Eve.

Beaconsfield, Lord. By T. P. O'CONNOR, M.P.   Cr. 8vo, cloth, 5s.

Bechstein (Ludwig).—As Pretty as Seven, and other German
Stories. With Additional Tales by the Brothers GRIMM, and 98 Illustrations by RICHTER. Square 8vo, cloth extra, 6s. 6d.: gilt edges, 7s. 6d.

Bennett (Arnold), Novels by.   Crown 8vo, cloth, gilt top, 6s. each.
The Grand Babylon Hotel: A Fantasia on Modern Themes. | Anna of the Five Towns.
Leonora.

The Gates of Wrath. Crown 8vo, cloth, 3s. 6d.

Bennett (W. C., LL.D.).—Songs for Sailors.   Post 8vo, cl. limp, 2s.

Bewick (Thomas) and his Pupils.   By AUSTIN DOBSON.   With 95
Illustrations. Square 8vo, cloth extra, 3s. 6d.

Bierce (Ambrose).—In the Midst of Life: Tales of Soldiers and
Civilians. Crown 8vo, cloth extra, 3s. 6d.; post 8vo, illustrated boards, 2s.

Bill Nye's Comic History of the United States.   With 146 Illustrations by F. OPPER. Crown 8vo, cloth extra, 3s. 6d.

Bindloss (Harold), Novels by.   Crown 8vo, cloth, gilt top, 6s. each.
A Sower of Wheat. | The Concession-Hunters. | The Mistress of Bonaventure.

Ainslie's Ju-Ju: A Romance of the Hinterland. Cr. 8vo, cloth, 3s. 6d.; picture cloth, flat back, 2s

Bodkin (M. McD., K.C.), Books by.
Dora Myrl, the Lady Detective. Crown 8vo, cloth, 3s. 6d.; picture cloth, flat back, 2s.
Shillelagh and Shamrock. Crown 8vo, cloth, 3s. 6d.

Bourget (Paul).—A Living Lie.   Translated by JOHN DE VILLIERS.
With special Preface for the English Edition. Crown 8vo, cloth, 3s. 6d.

Bourne (H. R. Fox), Books by.
English Merchants: Memoirs in Illustration of the Progress of British Commerce. With 32 Illus-
trations. Crown 8vo, cloth, 3s. 6d.
English Newspapers: Chapters in the History of Journalism. Two Vols., demy 8vo, cloth, 25s.
The Other Side of the Emin Pasha Relief Expedition. Crown 8vo, cloth, 6s.

Boyd.—A Versailles Christmas-tide.   By MARY STUART BOYD.   With
53 Illustrations by A. S. BOYD. Fcap. 4to. cloth gilt and gilt top, 6s.

Boyle (Frederick), Works by.   Post 8vo, illustrated bds., 2s. each.
Chronicles of No-Man's Land. | Camp Notes. | Savage Life.

Brand (John).—Observations on Popular Antiquities; chiefly
illustrating the Origin of our Vulgar Customs, Ceremonies, and Superstitions. With the Additions of Sir
HENRY ELLIS. Crown 8vo, cloth, 3s. 6d.

Brayshaw (J. Dodsworth).—Slum Silhouettes: Stories of London
Life. Crown 8vo, cloth, 3s. 6d.

Brewer's (Rev. Dr.) Dictionaries.   Crown 8vo, cloth, 3s. 6d. each.
The Reader's Handbook of Famous Names in Fiction, Allusions, References,
Proverbs, Plots, Stories, and Poems. A New Edition, Revised.
A Dictionary of Miracles: Imitative, Realistic, and Dogmatic.

Brewster (Sir David), Works by.   Post 8vo, cloth, 4s. 6d. each.
More Worlds than One: The Creed of the Philosopher and Hope of the Christian. With Plates.
The Martyrs of Science: GALILEO, TYCHO BRAHE, and KEPLER. With Portraits.
Letters on Natural Magic. With numerous Illustrations.

Bright (Florence).—A Girl Capitalist.   Cr. 8vo, cloth, gilt top, 6s.

Brillat-Savarin.—Gastronomy as a Fine Art.   Translated by
R. E. ANDERSON, M.A. Post 8vo, half-bound, 2s.

Bryden (H. A.).—An Exiled Scot: A Romance.   With a Frontispiece, by J. S. CROMPTON, R.I. Crown 8vo, cloth, 3s. 6d.

Brydges (Harold).—Uncle Sam at Home.   With 91 Illustrations.
Post 8vo, illustrated boards, 2s.; cloth limp, 2s. 6d.

Burton (Robert).—The Anatomy of Melancholy.   With Translations of the Quotations. Demy 8vo, cloth extra, 7s. 6d.
Melancholy Anatomised: An Abridgment of BURTON'S ANATOMY. Post 8vo, half-cl., 2s. 6d.

Buchanan (Robert), Poems and Novels by.

The Complete Poetical Works of Robert Buchanan. 2 vols., crown 8vo, buckram, with Portrait Frontispiece to each volume, 12s.

Crown 8vo, cloth, 3s. 6d. each ; post 8vo, illustrated boards, 2s. each.

The Shadow of the Sword.
A Child of Nature. With Frontispiece.
God and the Man. With 11 Illustrations by
Lady Kilpatrick.   [FRED. BARNARD.]
The Martyrdom of Madeline. With Frontispiece by A. W. COOPER.
Love Me for Ever. With Frontispiece.
Annan Water. | **Foxglove Manor.**
The New Abelard. | **Rachel Dene.**
Matt : A Story of a Caravan. With Frontispiece.
The Master of the Mine. With Frontispiece.
The Heir of Linne. | **Woman and the Man.**

Crown 8vo, cloth, 3s. 6d. each.

Red and White Heather. | **Andromeda :** An Idyll of the Great River.

The Shadow of the Sword. POPULAR EDITION, medium 8vo, 6d.

The Charlatan. By ROBERT BUCHANAN and HENRY MURRAY.   Crown 8vo, cloth, with a Frontispiece by T. H. ROBINSON, 3s. 6d. ; post 8vo, picture boards, 2s.

Burgess (Gelett) and Will Irwin.—The Picaroons : A San Fran-
cisco Night's Entertainment. Crown 8vo, cloth, 3s. 6d. [Jan

Caine (Hall), Novels by. Crown 8vo, cloth extra, 3s. 6d. each. ; post
8vo, illustrated boards, 2s. each ; cloth limp, 2s. 6d. each.
The Shadow of a Crime. | **A Son of Hagar.** | **The Deemster.**

Also LIBRARY EDITIONS of the three novels, set in new type, crown 8vo, bound uniform with **The Christian,** 6s. each ; and CHEAP POPULAR EDITIONS, medium 8vo, portrait-cover, 6d. each.— Also the FINE-PAPER EDITION of **The Deemster,** pott 8vo, cloth, gilt top, 2s. net ; leather, gilt edges, 3s. net.

Cameron (Commander V. Lovett).—The Cruise of the 'Black
Prince' Privateer. Post 8vo, picture boards, 2s.

Canada (Greater) : The Past, Present, and Future of the Canadian
North-West. By E. B. OSBORN, B.A. With a Map. Crown 8vo, cloth, 3s. 6d.

Captain Coignet, Soldier of the Empire : An Autobiography.
Edited by LOREDAN LARCHEY. Translated by Mrs. CAREY. With 100 Illustrations. Crown 8vo, cloth, 3s. 6d.

Carlyle (Thomas).—On the Choice of Books. Post 8vo, cl., 1s. 6d.

Carruth (Hayden).—The Adventures of Jones. With 17 Illustra
tions. Fcap. 8vo, cloth, 2s.

Chambers (Robert W.), Stories of Paris Life by.
The King in Yellow. Crown 8vo, cloth, 3s. 6d. ; fcap. 8vo, cloth limp, 2s. 6d.
In the Quarter. Fcap. 8vo, cloth, 2s. 6d.

Chapman's (George), Works. Vol. I., Plays Complete, including the
Doubtful Ones.—Vol. II., Poems and Minor Translations, with Essay by A. C. SWINBURNE.—Vol. III., Translations of the Iliad and Odyssey. Three Vols., crown 8vo, cloth, 3s. 6d. each.

Chapple (J. Mitchell).—The Minor Chord : The Story of a Prima
Donna. Crown 8vo, cloth, 3s. 6d.

Chaucer for Children : A Golden Key. By Mrs. H. R. HAWEIS. With
8 Coloured Plates and 30 Woodcuts. Crown 4to, cloth extra, 3s. 6d.
Chaucer for Schools. With the Story of his Times and his Work. By Mrs. H. R. HAWEIS A New Edition, revised. With a Frontispiece. Demy 8vo, cloth, 2s. 6d.

Chess, The Laws and Practice of. With an Analysis of the Open-
ings. By HOWARD STAUNTON. Edited by R. B. WORMALD. Crown 8vo, cloth, 5s.
The Minor Tactics of Chess : A Treatise on the Deployment of the Forces in obedience to Strategic Principle. By F. K. YOUNG and E. C. HOWELL. Long fcap. 8vo, cloth, 2s. 6d.
The Hastings Chess Tournament. Containing the Authorised Account of the 230 Games played Aug.-Sept., 1895. With Annotations by PILLSBURY, LASKER, TARRASCH, STEINITZ, SCHIFFERS, TEICHMANN, BARDELEBEN, BLACKBURNE, GUNSBERG, TINSLEY, MASON, and ALBIN ; Biographical Sketches of the Chess Masters, and 22 Portraits. Edited by H. F. CHESHIRE. Cheaper Edition. Crown 8vo, cloth, 5s.

Clare (Austin), Stories by.
For the Love of a Lass. Post 8vo, illustrated boards, 2s. ; cloth, 2s. 6d.
By the Rise of the River : Tales and Sketches in South Tynedale. Crown 8vo, cloth, 3s. 6d.
The Tideway. Crown 8vo, cloth, gilt top, 6s.

Clive (Mrs. Archer), Novels by.
Post 8vo, cloth, 3s. 6d. each ; picture boards, 2s. each.
Paul Ferroll. | **Why Paul Ferroll Killed his Wife.**

Clodd (Edward, F.R.A.S.).—Myths and Dreams. Cr. 8vo, 3s. 6d.

Coates (Anne).—Rie's Diary. Crown 8vo, cloth, 3s. 6d.

Cobban (J. Maclaren), Novels by.
The Cure of Souls. Post 8vo, illustrated boards, 2s.
The Red Sultan. Crown 8vo, cloth extra, 3s. 6d. ; post 8vo, illustrated boards, 2s.
The Burden of Isabel. Crown 8vo, cloth extra, 3s. 6d.

Collins (C. Allston).—The Bar Sinister. Post 8vo, boards, 2s.

Collins (John Churton, M.A.), Books by. Cr. 8vo, cl., 3s. 6d. each.
Illustrations of Tennyson.
Jonathan Swift. A Biographical and Critical Study.

Collins (Mortimer and Frances), Novels by.
Crown 8vo, cloth extra, 3s. 6d. each ; post 8vo, illustrated boards, 2s. each.

From Midnight to Midnight.	Blacksmith and Scholar.
You Play me False.	The Village Comedy.

Post 8vo, illustrated boards, 2s. each.

Transmigration.	Sweet Anne Page.	Frances.
A Fight with Fortune.	Sweet and Twenty.	

Collins (Wilkie), Novels by.
Crown 8vo, cloth extra, many Illustrated, 3s. 6d. each ; post 8vo, picture boards, 2s. each ; cloth limp, 2s. 6d. each.

*Antonina.	My Miscellanies.	Jezebel's Daughter.
*Basil.	Armadale.	The Black Robe.
*Hide and Seek.	Poor Miss Finch.	Heart and Science.
*The Woman in White.	Miss or Mrs.?	'I Say No.'
*The Moonstone.	The New Magdalen.	A Rogue's Life.
*Man and Wife.	The Frozen Deep.	The Evil Genius.
*The Dead Secret.	The Law and the Lady.	Little Novels.
After Dark.	The Two Destinies.	The Legacy of Cain.
The Queen of Hearts.	The Haunted Hotel.	Blind Love.
No Name.	The Fallen Leaves.	

₌ Marked * have been reset in new type, in uniform style.

POPULAR EDITIONS, medium 8vo, 6d. each.

The Moonstone.	Antonina.	The Dead Secret.
The Woman in White.	The New Magdalen.	Man and Wife.	Armadale.

The Woman in White. LARGE TYPE, FINE PAPER EDITION. Pott 8vo, cloth, gilt top, 2s. net ; leather, gilt edges, 3s. net.

Colman's (George) Humorous Works: 'Broad Grins,' 'My Night-gown and Slippers,' &c. With Life and Frontispiece. Crown 8vo, cloth extra, 3s. 6d.

Colquhoun (M. J.).—Every Inch a Soldier. Crown 8vo, cloth, 3s. 6d. ; post 8vo, illustrated boards, 2s.

Colt-breaking, Hints on. By W. M. HUTCHISON. Cr. 8vo, cl., 3s. 6d.

Compton (Herbert), Novels by.
The Inimitable Mrs. Massingham. Crown 8vo, cloth, 3s. 6d.
The Wilful Way. Crown 8vo, cloth, gilt top, 6s.

Convalescent Cookery. By CATHERINE RYAN. Cr. 8vo, 1s. ; cl., 1s. 6d.

Cooper (Edward H.).—Geoffory Hamilton. Cr. 8vo, cloth, 3s. 6d.

Cornish (J. F.).—Sour Grapes : A Novel. Cr. 8vo, cloth, gilt top, 6s.

Cornwall.—Popular Romances of the West of England; or, The Drolls, Traditions, and Superstitions of Old Cornwall. Collected by ROBERT HUNT, F.R.S. With two Steel Plates by GEORGE CRUIKSHANK. Crown 8vo, cloth, 7s. 6d.

Cotes (V. Cecil).—Two Girls on a Barge. With 44 Illustrations by F. H. TOWNSEND. Crown 8vo, cloth extra, 3s. 6d. ; post 8vo, cloth, 2s. 6d.

Craddock (C. Egbert), Stories by.
The Prophet of the Great Smoky Mountains. Crown 8vo, cloth, 3s. 6d. ; post 8vo, illustrated boards, 2s.
His Vanished Star. Crown 8vo, cloth, 3s. 6d.

Crellin (H. N.).—Romances of the Old Seraglio. With 28 Illustrations by S. L. WOOD. Crown 8vo, cloth, 3s. 6d.

Cresswell (Henry).—A Lady of Misrule. Cr. 8vo, cloth, gilt top, 6s.

Crim (Matt.).—Adventures of a Fair Rebel. Crown 8vo, cloth extra, with a Frontispiece by DAN. BEARD, 3s. 6d. ; post 8vo, illustrated boards, 2s.

Crockett (S. R.) and others. — Tales of Our Coast. By S. R. CROCKETT, GILBERT PARKER, HAROLD FREDERIC, 'Q.,' and W. CLARK RUSSELL. With 2 Illustrations by FRANK BRANGWYN. Crown 8vo, cloth, 3s. 6d.

Croker (Mrs. B. M.), Novels by. Crown 8vo, cloth extra, 3s. 6d. each; post 8vo, illustrated boards, 2s. each; cloth limp, 2s. 6d. each.

Pretty Miss Neville.	Interference.	Village Tales & Jungle
Proper Pride.	A Family Likeness.	Tragedies.
A Bird of Passage.	A Third Person.	The Real Lady Hilda.
Diana Barrington.	Mr. Jervis.	Married or Single?
Two Masters.		

Crown 8vo, cloth extra, 3s. 6d. each.

Some One Else.	Miss Balmaine's Past.	Beyond the Pale.
In the Kingdom of Kerry.	Jason, &c.	Infatuation.
Terence. With 6 Illustrations by SIDNEY PAGET.		

'To Let,' &c. Post 8vo, picture boards, 2s.; cloth limp, 2s. 6d.
The Cat's-paw. With 12 Illustrations by FRED. PEGRAM. Crown 8vo, cloth, gilt top, 6s
Diana Barrington. POPULAR EDITION, medium 8vo, 6d.

Cruikshank's Comic Almanack. Complete in TWO SERIES: The FIRST, from 1835 to 1843; the SECOND, from 1844 to 1853. A Gathering of the Best Humour of THACKERAY, HOOD, MAYHEW, ALBERT SMITH, A'BECKETT, ROBERT BROUGH, &c. With numerous Steel Engravings and Woodcuts by GEORGE CRUIKSHANK, HINE, LANDELLS, &c. Two Vols., crown 8vo, cloth gilt, 7s. 6d. each.
The Life of George Cruikshank. By BLANCHARD JERROLD. With 84 Illustrations and a Bibliography. Crown 8vo, cloth extra, 3s. 6d.

Cumming (C. F. Gordon), Works by. Large cr. 8vo, cloth, 6s. each.
In the Hebrides. With an Autotype Frontispiece and 23 Illustrations.
In the Himalayas and on the Indian Plains. With 42 Illustrations.
Two Happy Years in Ceylon. With 29 Illustrations.
Via Cornwall to Egypt. With a Photogravure Frontispiece.

Cussans (John E.).—A Handbook of Heraldry; with Instructions for Tracing Pedigrees and Deciphering Ancient MSS., &c. Fourth Edition, revised, with 408 Woodcuts and 2 Coloured Plates. Crown 8vo, cloth extra, 6s.

Daudet (Alphonse).—The Evangelist; or, Port Salvation. Crown 8vo, cloth extra, 3s. 6d.; post 8vo, illustrated boards, 2s.

Davenant (Francis, M.A.).—Hints for Parents on the Choice of a Profession for their Sons when Starting in Life. Crown 8vo, cloth, 1s. 6d.

Davidson (Hugh Coleman).—Mr. Sadler's Daughters. With a Frontispiece by STANLEY WOOD. Crown 8vo, cloth extra, 3s. 6d.

Davies (Dr. N. E. Yorke-), Works by. Cr. 8vo, 1s. ea.; cl., 1s. 6d. ea.
One Thousand Medical Maxims and Surgical Hints.
Nursery Hints: A Mother's Guide in Health and Disease.
Foods for the Fat: The Dietetic Cure of Corpulency and of Gout.
Aids to Long Life. Crown 8vo, 2s.; cloth limp, 2s. 6d.

Davies' (Sir John) Complete Poetical Works. Collected and Edited, with Introduction and Notes, by Rev. A. B. GROSART, D.D. Two Vols., crown 8vo, cloth, 3s. 6d. each.

De Guerin (Maurice), The Journal of. Edited by G. S. TREBUTIEN. With a Memoir by SAINTE-BEUVE. Translated from the 20th French Edition by JESSIE P. FROTHINGHAM. Fcap. 8vo, half-bound, 2s. 6d.

De Maistre (Xavier).—A Journey Round my Room. Translated by HENRY ATTWELL. Post 8vo, cloth limp, 2s. 6d.

Derby (The): The Blue Ribbon of the Turf. With Brief Accounts of THE OAKS. By LOUIS HENRY CURZON. Crown 8vo, cloth limp, 2s. 6d.

Dewar (T. R.).—A Ramble Round the Globe. With 220 Illustrations. Crown 8vo, cloth extra, 7s. 6d.

De Windt (Harry), Books by.
Through the Gold-Fields of Alaska to Bering Straits. With Map and 33 full-page Illustrations. Cheaper Issue. Demy 8vo, cloth, 6s.
True Tales of Travel and Adventure. Crown 8vo, cloth, 3s. 6d.

Dickens (Charles), About England with. By ALFRED RIMMER. With 57 Illustrations by C. A. VANDERHOOF and the AUTHOR. Square 8vo, cloth, 3s. 6d.
Christmas Days with Boz. By PERCY FITZGERALD. With Illustrations in Facsimile of Old Prints. Crown 8vo, cloth, 3s. 6d. [Shortly.

Dictionaries.
The Reader's Handbook of Famous Names in Fiction, Allusions, References, Proverbs, Plots, Stories, and Poems. By Rev. E. C. BREWER, LL.D. A New Edition, Revised. Crown 8vo, cloth, 3s. 6d.
A Dictionary of Miracles: Imitative, Realistic, and Dogmatic. By the Rev. E. C. BREWER, LL.D. Crown 8vo, cloth, 3s. 6d.
Familiar Short Sayings of Great Men. With Historical and Explanatory Notes by SAMUEL A. BENT, A.M. Crown 8vo, cloth extra, 7s. 6d.
The Slang Dictionary: Etymological, Historical, and Anecdotal. Crown 8vo, cloth, 6s. 6d.
Words, Facts, and Phrases: A Dictionary of Curious, Quaint, and Out-of-the-Way Matters. By ELIEZER EDWARDS. Crown 8vo, cloth extra, 3s. 6d.

Dilke (Rt. Hon. Sir Charles, Bart., M.P.).—The British Empire.
Crown 8vo, buckram, 3s. 6d.

Dobson (Austin), Works by.
Thomas Bewick and his Pupils. With 95 Illustrations. Square 8vo, cloth, 3s. 6d.
Four Frenchwomen. With Four Portraits. Crown 8vo, buckram, gilt top, 6s.
Eighteenth Century Vignettes. IN THREE SERIES. Crown 8vo, buckram, 6s. each.
A Paladin of Philanthropy, and other Papers. With 2 Illusts. Cr. 8vo, buckram, 6s.
Side-walk Studies. With 5 Illustrations. SECOND EDITION. Crown 8vo, buckram, gilt top, 6s.

Dobson (W. T.).—Poetical Ingenuities and Eccentricities. Post
8vo, cloth limp, 2s. 6d.

Donovan (Dick), Detective Stories by.
Post 8vo, illustrated boards, 2s. each ; cloth limp, 2s. 6d. each.

The Man-Hunter.	Wanted!	Suspicion Aroused. Riddles Read.
Caught at Last.	Tracked to Doom.	A Detective's Triumphs.
Tracked and Taken. Link by Link.		In the Grip of the Law.
Who Poisoned Hetty Duncan?		From Information Received.

Crown 8vo, cloth extra, 3s. 6d. each : post 8vo, illustrated boards, 2s. each ; cloth, 2s. 6d. each.
The Man from Manchester. With 23 Illustrations.
The Mystery of Jamaica Terrace. | The Chronicles of Michael Danevitch.
Crown 8vo, cloth, 3s. 6d. each.
The Records of Vincent Trill, of the Detective Service.—Also picture cloth, flat back, 2s.
The Adventures of Tyler Tatlock, Private Detective.
Deacon Brodie ; or, Behind the Mask. | Tales of Terror.
Dark Deeds. Crown 8vo, picture cloth, flat back, 2s.

Dowling (Richard).—Old Corcoran's Money. Crown 8vo, cl., 3s. 6d.

Doyle (A. Conan).—The Firm of Girdlestone. Cr. 8vo, cl., 3s. 6d.

Dramatists, The Old. Cr. 8vo, cl. ex., with Portraits, 3s. 6d. per Vol.
Ben Jonson's Works. With Notes, Critical and Explanatory, and a Biographical Memoir by
WILLIAM GIFFORD. Edited by Colonel CUNNINGHAM. Three Vols.
Chapman's Works. Three Vols. Vol. I. contains the Plays complete ; Vol. II., Poems and Minor
Translations, with an Essay by A. C. SWINBURNE ; Vol. III., Translations of the Iliad and Odyssey.
Marlowe's Works. Edited, with Notes, by Colonel CUNNINGHAM. One Vol.
Massinger's Plays. From GIFFORD'S Text. Edited by Colonel CUNNINGHAM. One Vol.

Dublin Castle and Dublin Society, Recollections of. By A
NATIVE. Crown 8vo, cloth, gilt top. 6s.

Duncan (Sara Jeannette: Mrs. EVERARD COTES), Books by.
Crown 8vo, cloth extra, 7s. 6d. each.
A Social Departure. With 111 Illustrations by F. H. TOWNSEND.
An American Girl in London. With 80 Illustrations by F. H. TOWNSEND.
The Simple Adventures of a Memsahib. With 37 Illustrations by F. H. TOWNSEND.
Crown 8vo, cloth extra, 3s. 6d. each.
A Daughter of To-Day. | Vernon's Aunt. With 47 Illustrations by HAL HURST.

Dutt (Romesh C.).—England and India: A Record of Progress
during One Hundred Years. Crown 8vo, cloth, 2s.

Early English Poets. Edited, with Introductions and Annotations,
by Rev. A. B. GROSART, D.D. Crown 8vo, cloth boards, 3s. 6d. per Volume.
Fletcher's (Giles) Complete Poems. One Vol.
Davies' (Sir John) Complete Poetical Works. Two Vols.
Sidney's (Sir Philip) Complete Poetical Works. Three Vols.

Edgcumbe (Sir E. R. Pearce).—Zephyrus: A Holiday in Brazil
and on the River Plate. With 41 Illustrations. Crown 8vo, cloth extra, 5s.

Edwardes (Mrs. Annie), Novels by.
A Point of Honour. Post 8vo, illustrated boards, 2s. | A Plaster Saint. Cr. 8vo, cl., 3s. 6d.
Archie Lovell. Crown 8vo, cloth, 3s. 6d. ; illustrated boards, 2s.

Edwards (Eliezer).—Words, Facts, and Phrases: A Dictionary
of Curious, Quaint, and Out-of-the-Way Matters. Cheaper Edition. Crown 8vo, cloth, 3s. 6d.

Egerton (Rev. J. C., M.A.). — Sussex Folk and Sussex Ways.
With Introduction by Rev. Dr. H. WACE, and Four Illustrations. Crown 8vo, cloth extra, 5s.

Eggleston (Edward).—Roxy: A Novel. Post 8vo, illust. boards, 2s.

Englishman (An) in Paris. Notes and Recollections during the
Reign of Louis Philippe and the Empire. Crown 8vo, cloth, 3s. 6d.

Englishman's House, The: A Practical Guide for Selecting or Build-
ing a House. By C. J. RICHARDSON. Coloured Frontispiece and 534 Illusts. Cr. 8vo, cloth, 3s. 6d.

Eyes, Our: How to Preserve Them. By JOHN BROWNING. Cr. 8vo, 1s.

Familiar Short Sayings of Great Men. By SAMUEL ARTHUR BENT,
A.M. Fifth Edition, Revised and Enlarged. Crown 8vo, cloth extra, 7s. 6d.

Faraday (Michael), Works by. Post 8vo, cloth extra, 4s. 6d. each.
   The Chemical History of a Candle: Lectures delivered before a Juvenile Audience. Edited by WILLIAM CROOKES, F.C.S. With numerous Illustrations.
   On the Various Forces of Nature, and their Relations to each other. Edited by WILLIAM CROOKES, F.C.S. With Illustrations.

Farrer (J. Anson).—War: Three Essays. Crown 8vo, cloth, 1s. 6d.

Fenn (G. Manville), Novels by.
   Crown 8vo, cloth extra, 3s. 6d. each; post 8vo, illustrated boards, 2s. each.
   The New Mistress. | Witness to the Deed. | The Tiger Lily. | The White Virgin.

Crown 8vo, cloth 3s. 6d. each.

A Woman Worth Winning.	Double Cunning.	The Story of Antony Grace
Cursed by a Fortune.	A Fluttered Dovecote.	The Man with a Shadow.
The Case of Ailsa Gray.	King of the Castle.	One Maid's Mischief.
Commodore Junk.	The Master of the Cere-	This Man's Wife.
Black Blood.	monies.	In Jeopardy.

Crown 8vo, cloth, gilt top, 6s. each.
   The Bag of Diamonds, and Three Bits of Paste.
   Running Amok: a Story of Adventure.
   The Cankerworm: being Episodes of a Woman's Life. | **Black Shadows.**
   A Crimson Crime. Crown 8vo, cloth, gilt top, 6s.; picture cloth, flat back, 2s.

Fiction, A Catalogue of, with Descriptive Notices and Reviews of over NINE HUNDRED NOVELS, will be sent free by Messrs. CHATTO & WINDUS upon application.

Fin-Bec.—The Cupboard Papers: Observations on the Art of Living and Dining. Post 8vo, cloth limp, 2s. 6d.

Firework-Making, The Complete Art of; or, The Pyrotechnist's Treasury. By THOMAS KENTISH. With 267 Illustrations. Crown 8vo, cloth, 3s. 6d.

First Book, My. By WALTER BESANT, JAMES PAYN, W. CLARK RUS-SELL, GRANT ALLEN, HALL CAINE, GEORGE R. SIMS, RUDYARD KIPLING, A. CONAN DOYLE, M. E. BRADDON, F. W. ROBINSON, H. RIDER HAGGARD, R. M. BALLANTYNE, I. ZANGWILL, MORLEY ROBERTS, D. CHRISTIE MURRAY, MARY CORELLI, J. K. JEROME, JOHN STRANGE WINTER, BRET HARTE, 'Q.,' ROBERT BUCHANAN, and R. L. STEVENSON. With a Prefatory Story by JEROME K. JEROME, and 185 Illustrations. A New Edition. Small demy 8vo, art linen, 3s. 6d.

Fitzgerald (Percy), Works by.
   Little Essays: Passages from the Letters of CHARLES LAMB. Post 8vo, cloth, 2s. 6d.
   Fatal Zero. Crown 8vo, cloth extra, 3s. 6d.; post 8vo, illustrated boards, 2s.

Post 8vo, illustrated boards, 2s. each.

Bella Donna.	The Lady of Brantome.	The Second Mrs. Tillotson.
Polly.	Never Forgotten.	Seventy-five Brooke Street.

   Sir Henry Irving: Twenty Years at the Lyceum. With Portrait. Crown 8vo, cloth, 1s. 6d.
   Christmas Days with Boz. With Illustrations in Facsimile of Old Prints. Crown 8vo, cloth, 3s. 6d.   [Shortly.

Flammarion (Camille), Works by.
   Popular Astronomy: A General Description of the Heavens. Translated by J. ELLARD GORE, F.R.A.S. With Three Plates and 288 Illustrations. Medium 8vo, cloth, 10s. 6d.
   Urania: A Romance. With 87 Illustrations. Crown 8vo, cloth extra, 5s.

Fletcher's (Giles, B.D.) Complete Poems: Christ's Victorie in Heaven, Christ's Victorie on Earth, Christ's Triumph over Death, and Minor Poems. With Notes by Rev. A. B. GROSART. D.D. Crown 8vo, cloth boards, 3s. 6d.

Forbes (Archibald).—The Life of Napoleon III. With Photo-gravure Frontispiece and Thirty-six full-page Illustrations. Cheaper Issue. Demy 8vo, cloth, 6s.

Forbes (Hon. Mrs. Walter R. D.).—Dumb. Crown 8vo, cl., 3s. 6d.

Francillon (R. E.), Novels by.
   Crown 8vo, cloth extra, 3s. 6d. each; post 8vo, illustrated boards, 2s. each.
   One by One. | A Real Queen. | A Dog and his Shadow. | Ropes of Sand. Illust.
   Post 8vo, illustrated boards, 2s. each.
   Queen Cophetua. | Olympia. | Romances of the Law. | King or Knave?
   Jack Doyle's Daughter. Crown 8vo, cloth, 3s. 6d.

Frederic (Harold), Novels by. Post 8vo, cloth extra, 3s. 6d. each; illustrated boards, 2s. each.
   Seth's Brother's Wife. | The Lawton Girl.

Fry's (Herbert) Royal Guide to the London Charities, 1904. Edited by JOHN LANE. Published Annually. Crown 8vo, cloth, 1s. 6d.

Gardening Books. Post 8vo, 1s. each; cloth limp, 1s. 6d. each.
   A Year's Work in Garden and Greenhouse. By GEORGE GLENNY.
   Household Horticulture. By TOM and JANE JERROLD. Illustrated.
   The Garden that Paid the Rent. By TOM JERROLD.

Gaulot (Paul), Books by:
   The Red Shirts: A Tale of "The Terror." Translated by JOHN DE VILLIERS. With a Fron-tispiece by STANLEY WOOD. Crown 8vo, cloth, 3s. 6d.; picture cloth, flat back, 2s.
   Love and Lovers of the Past. Translated by CHARLES LAROCHE, M.A. With a Frontis-piece. Crown 8vo, cloth, gilt top, 6s.

Gentleman's Magazine, The. 1s. Monthly. Contains Stories, Articles upon Literature, Science, Biography, and Art, and 'Table Talk' by SYLVANUS URBAN.
*** *Bound Volumes for recent years kept in stock, 8s. 6d. each. Cases for binding, 2s. each.*

German Popular Stories. Collected by the Brothers GRIMM and Translated by EDGAR TAYLOR. With Introduction by JOHN RUSKIN, and 22 Steel Plates after GEORGE CRUIKSHANK. Square 8vo, cloth, 6s. 6d. ; gilt edges, 7s. 6d.

Gibbon (Chas.), Novels by. Cr. 8vo, cl., 3s. 6d. ea.; post 8vo, bds., 2s. ea.

Robin Gray. With Frontispiece.	**Loving a Dream.** | **The Braes of Yarrow.**
The Golden Shaft. With Frontispiece.	**Of High Degree.**
The Flower of the Forest.	**Queen of the Meadow.**

Post 8vo, illustrated boards, 2s. each.

The Dead Heart.	**In Pastures Green.**	**In Honour Bound.**
For Lack of Gold.	**In Love and War.**	**Heart's Delight.**
What Will the World Say?	**A Heart's Problem.**	**Blood-Money.**
For the King.	**By Mead and Stream.**	
A Hard Knot.	**Fancy Free.**	

Gibney (Somerville).—Sentenced ! Crown 8vo, cloth, 1s. 6d.

Gilbert's (W. S.) Original Plays. In 3 Series, post 8vo, 2s. 6d. each.
The FIRST SERIES contains: The Wicked World—Pygmalion and Galatea—Charity—The Princess—The Palace of Truth—Trial by Jury—Iolanthe.
The SECOND SERIES: Broken Hearts—Engaged—Sweethearts—Gretchen—Dan'l Druce—Tom Cobb—H.M.S. 'Pinafore'—The Sorcerer—The Pirates of Penzance.
The THIRD SERIES: Comedy and Tragedy—Foggerty's Fairy—Rosencrantz and Guildenstern—Patience—Princess Ida—The Mikado—Ruddigore—The Yeomen of the Guard—The Gondoliers—The Mountebanks—Utopia.

Eight Original Comic Operas written by W. S. GILBERT. Two Series, demy 8vo, cloth, 2s. 6d. each. The FIRST SERIES contains: The Sorcerer—H.M.S. 'Pinafore'—The Pirates of Penzance—Iolanthe—Patience—Princess Ida—The Mikado—Trial by Jury.
The SECOND SERIES contains: The Gondoliers—The Grand Duke—The Yeomen of the Guard—His Excellency—Utopia, Limited—Ruddigore—The Mountebanks—Haste to the Wedding.
The Gilbert and Sullivan Birthday Book: Quotations for Every Day in the Year, selected from Plays by W. S. GILBERT set to Music by Sir A. SULLIVAN. Compiled by ALEX. WATSON. Royal 16mo, Japanese leather, 2s. 6d.

Gilbert (William). — James Duke, Costermonger. Post 8vo, illustrated boards, 2s.

Gissing (Algernon), Novels by. Crown 8vo, cloth, gilt top, 6s. each.

A Secret of the North Sea.	**The Wealth of Mallerstang.**
Knitters in the Sun.	**An Angel's Portion.**

Glanville (Ernest), Novels by.
Crown 8vo, cloth extra, 3s. 6d. each ; post 8vo, illustrated boards, 2s. each.
The Lost Heiress: A Tale of Love, Battle, and Adventure. With Two Illustrations by H. NISBET.
The Fossicker: A Romance of Mashonaland. With Two Illustrations by HUME NISBET.
A Fair Colonist. With a Frontispiece by STANLEY WOOD.

The Golden Rock. With a Frontispiece by STANLEY WOOD. Crown 8vo, cloth extra, 3s. 6d.
Kloof Yarns. Crown 8vo cloth, 1s. 6d.
Tales from the Veld. With Twelve Illustrations by M. NISBET. Crown 8vo, cloth, 3s. 6d.
Max Thornton. With 8 Illustrations by J. S. CROMPTON, R.I. Large crown 8vo, cloth, gilt edges, 5s. ; cloth, gilt top, 6s.

Glenny (George).—A Year's Work in Garden and Greenhouse:
Practical Advice as to the Management of the Flower, Fruit, and Frame Garden. Post 8vo, 1s. ; cloth, 1s. 6d.

Godwin (William).—Lives of the Necromancers. Post 8vo, cl., 2s.

Golden Treasury of Thought, The: A Dictionary of Quotations from the Best Authors. By THEODORE TAYLOR. Crown 8vo, cloth, 3s. 6d.

Goodman (E. J.).—The Fate of Herbert Wayne. Cr. 8vo, 3s. 6d.

Gore (J. Ellard, F.R.A.S.).—The Stellar Heavens: an Introduction to the Study of the Stars and Nebulæ. Crown 8vo, cloth, 2s. net.

Grace (Alfred A.).—Tales of a Dying Race. Cr. 8vo, cloth, 3s. 6d.

Greeks and Romans, The Life of the, described from Antique Monuments. By ERNST GUHL and W. KONER. Edited by Dr. F. HUEFFER. With 545 Illustrations. Large crown 8vo, cloth extra, 7s. 6d.

Greenwood (James: "The Amateur Casual").—The Prisoner in the Dock: My Four Years' Daily Experiences in the London Police Courts. Cr. 8vo, cl., 3s. 6d.

Grey (Sir George).—The Romance of a Proconsul: Being the Personal Life and Memoirs of Sir GEORGE GREY, K.C.B. By JAMES MILNE. With Portrait. SECOND EDITION. Crown 8vo, buckram, 6s.

Griffith (Cecil).—Corinthia Marazion: A Novel. Crown 8vo, cloth extra, 3s. 6d.

Gunter (A. Clavering, Author of 'Mr. Barnes of New York').— A Florida Enchantment. Crown 8vo, cloth, 3s. 6d.

Guttenberg (Violet), Novels by.
Neither Jew nor Greek. | The Power of the Palmist.

Hair, The: Its Treatment in Health, Weakness, and Disease. Translated from the German of Dr. J. PINCUS. Crown 8vo, 1s. ; cloth, 1s. 6d.

Hake (Dr. Thomas Gordon), Poems by. Cr. 8vo, cl. ex., 6s. each.
New Symbols. | Legends of the Morrow. | The Serpent Play.
Maiden Ecstasy. Small 4to, cloth extra. 8s.

Halifax (C.).—Dr. Rumsey's Patient. By Mrs. L. T. MEADE and CLIFFORD HALIFAX, M.D. Crown 8vo, cloth, 3s. 6d.

Hall (Mrs. S. C.).—Sketches of Irish Character. With numerous Illustrations on Steel and Wood by MACLISE, GILBERT, HARVEY, and GEORGE CRUIKSHANK. Small demy 8vo, cloth extra, 7s. 6d.

Hall (Owen), Novels by.
The Track of a Storm. Crown 8vo, cloth, 3s. 6d. ; picture cloth, flat back, 2s.
Jetsam. Crown 8vo, cloth, 3s. 6d.

Crown 8vo, cloth, gilt top, 6s. each.
Eureka. | Hernando.

Halliday (Andrew).—Every-day Papers. Post 8vo, picture bds., 2s.

Hamilton (Cosmo), Stories by. Crown 8vo, cloth gilt, 3s. 6d. each.
The Glamour of the Impossible. | Through a Keyhole.
⁎ The two stories may also be had bound together in one volume, crown 8vo, cloth, 3s. 6d.

Harte's (Bret) Collected Works. Revised by the Author. LIBRARY EDITION, in Ten Volumes, crown 8vo, cloth extra, 6s. each.
Vol. I. COMPLETE POETICAL AND DRAMATIC WORKS. With Steel-plate Portrait.
  „ II. THE LUCK OF ROARING CAMP—BOHEMIAN PAPERS—AMERICAN LEGEND.
  „ III. TALES OF THE ARGONAUTS—EASTERN SKETCHES.
  „ IV. GABRIEL CONROY. | Vol. V. STORIES—CONDENSED NOVELS, &c.
  „ VI. TALES OF THE PACIFIC SLOPE.
  „ VII. TALES OF THE PACIFIC SLOPE—II. With Portrait by JOHN PETTIE, R.A.
  „ VIII. TALES OF THE PINE AND THE CYPRESS.
  „ IX. BUCKEYE AND CHAPPAREL.
  „ X. TALES OF TRAIL AND TOWN, &c.

Bret Harte's Choice Works, in Prose and Verse. With Portrait of the Author and 40 Illustrations. Crown 8vo, cloth, 3s. 6d.
Bret Harte's Poetical Works, including "Some Later Verses." Crown 8vo, buckram, 4s. 6d.
Some Later Verses. Crown 8vo, linen gilt, 5s.
In a Hollow of the Hills. Crown 8vo, picture cloth, flat back, 2s.
Condensed Novels. (The Two Series in One Volume.) Pott 8vo, cloth, gilt top, 2s. net ; leather, gilt edges, 3s. net.

Crown 8vo, cloth, 6s. each.
On the Old Trail. | From Sandhill to Pine.
Under the Redwoods. | Stories in Light and Shadow.
Mr. Jack Hamlin's Mediation.

Crown 8vo, cloth extra, 3s. 6d. each ; post 8vo, picture boards, 2s. each.
Gabriel Conroy.
A Waif of the Plains. With 60 Illustrations by STANLEY L. WOOD.
A Ward of the Golden Gate. With 59 Illustrations by STANLEY L. WOOD.

Crown 8vo, cloth extra, 3s. 6d. each.
Susy: A Novel. With Frontispiece and Vignette by J. A. CHRISTIE.
Sally Dows, &c. With 47 Illustrations by W. D. ALMOND and others.
The Bell-Ringer of Angel's, &c. With 39 Illustrations by DUDLEY HARDY and others.
Clarence: A Story of the American War. With Eight Illustrations by A. JULE GOODMAN.
Barker's Luck, &c. With 39 Illustrations by A. FORESTIER, PAUL HARDY, &c.
Devil's Ford, &c. With a Frontispiece by W. H. OVEREND.
The Crusade of the "Excelsior." With a Frontispiece by J. BERNARD PARTRIDGE.
Three Partners ; or, The Big Strike on Heavy Tree Hill. With 8 Illustrations by J. GULICH.
Tales of Trail and Town. With Frontispiece by G. P. JACOMB-HOOD.
New Condensed Novels ; Burlesques.

Crown 8vo, cloth, 3s. 6d. each ; picture cloth, flat backs. 2s. each.
The Luck of Roaring Camp, and Sensation Novels Condensed.
A Sappho of Green Springs. | Colonel Starbottle's Client.
A Protegee of Jack Hamlin's. With numerous Illustrations.

Post 8vo, illustrated boards, 2s. each.
An Heiress of Red Dog. | The Luck of Roaring Camp. | Californian Stories.

Post 8vo, illustrated boards, 2s. each ; cloth, 2s. 6d. each.
Flip. | Maruja. | A Phyllis of the Sierras.

Handwriting, The Philosophy of. With over 100 Facsimiles and Explanatory Text. By DON FELIX DE SALAMANCA. Post 8vo, half-cloth, 2s. 6d.

Hanky-Panky: Easy and Difficult Tricks, White Magic, Sleight of Hand, &c. Edited by W. H. CREMER. With 200 Illustrations. Crown 8vo, cloth extra, 4s. 6d.

Hardy (Rev. E. J., Author of 'How to be Happy though Married').—Love, Courtship, and Marriage. Crown 8vo, cloth, 3s. 6d.

Hardy (Iza Duffus), Novels by.
Crown 8vo, cloth, gilt top, 6s. each.
The Lesser Evil. | Man, Woman, and Fate.
A Butterfly: Her Friends and her Fortunes.

Hardy (Thomas).—Under the Greenwood Tree. Post 8vo, cloth extra, 3s. 6d.; illustrated boards, 2s.; cloth limp, 2s. 6d.—Also the FINE PAPER EDITION, pott 8vo, cloth, gilt top, 2s. net; leather, gilt edges, 3s. net.

Haweis (Mrs. H. R.), Books by.
The Art of Beauty. With Coloured Frontispiece and 91 Illustrations. Square 8vo, cloth bds., 6s.
The Art of Decoration. With Coloured Frontispiece and 74 Illustrations. Sq. 8vo, cloth bds., 6s.
The Art of Dress. With 32 Illustrations. Post 8vo, 1s.; cloth, 1s. 6d.
Chaucer for Schools. With the Story of his Times and his Work. A New Edition, revised. With a Frontispiece. Demy 8vo, cloth, 2s. 6d.
Chaucer for Children. With 38 Illustrations (8 Coloured). Crown 4to, cloth extra, 3s. 6d.

Haweis (Rev. H. R., M.A.).—American Humorists: WASHINGTON IRVING, OLIVER WENDELL HOLMES, JAMES RUSSELL LOWELL, ARTEMUS WARD, MARK TWAIN, and BRET HARTE. Crown 8vo, cloth, 6s.

Hawthorne (Julian), Novels by.
Crown 8vo, cloth extra, 3s. 6d. each; post 8vo, illustrated boards, 2s. each.
Garth. | Ellice Quentin. | Beatrix Randolph. With Four Illusts.
Fortune's Fool. | Dust. Four Illusts. | David Poindexter's Disappearance.
| | The Spectre of the Camera.

Post 8vo, illustrated boards, 2s. each.
Miss Cadogna. | Love—or a Name.
Sebastian Strome. Crown 8vo, cloth, 3s. 6d.

Heckethorn (C. W.), Books by. Crown 8vo, cloth, gilt top, 6s. each.
London Souvenirs. | London Memories: Social, Historical, and Topographical.

Helps (Sir Arthur), Books by. Post 8vo, cloth limp, 2s. 6d. each.
Animals and their Masters. | Social Pressure.
Ivan de Biron: A Novel. Crown 8vo, cloth extra, 3s. 6d.; post 8vo, illustrated boards, 2s.

Henderson (Isaac).—Agatha Page: A Novel. Cr. 8vo, cl., 3s. 6d.

Henty (G. A.), Novels by.
Rujub, the Juggler. Post 8vo, cloth, 3s. 6d.; illustrated boards, 2s.
Colonel Thorndyke's Secret. With a Frontispiece by STANLEY L. WOOD. Small demy 8vo, cloth, gilt edges, 5s.

Crown 8vo, cloth, 3s. 6d. each.
The Queen's Cup. | Dorothy's Double.

Herman (Henry).—A Leading Lady. Post 8vo, cloth, 2s. 6d.

Hertzka (Dr. Theodor).—Freeland: A Social Anticipation. Translated by ARTHUR RANSOM. Crown 8vo, cloth extra, 6s.

Hesse-Wartegg (Chevalier Ernst von).— Tunis: The Land and the People. With 22 Illustrations. Crown 8vo, cloth extra, 3s. 6d.

Hill (Headon).—Zambra the Detective. Crown 8vo, cloth, 3s. 6d.; post 8vo, picture boards, 2s.

Hill (John), Works by.
Treason-Felony. Post 8vo, boards, 2s. | The Common Ancestor. Cr. 8vo, cloth, 3s. 6d.

Hinkson (H. A.), Novels by. Crown 8vo, cloth, gilt top, 6s. each.
Fan Fitzgerald. | Silk and Steel.

Hoey (Mrs. Cashel).—The Lover's Creed. Post 8vo, boards, 2s.

Holiday, Where to go for a. By E. P. SHOLL, Sir H. MAXWELL, Bart., M.P., JOHN WATSON, JANE BARLOW, MARY LOVETT CAMERON, JUSTIN H. McCARTHY, PAUL LANGE, J. W. GRAHAM, J. H. SALTER, PHŒBE ALLEN, S. J. BECKETT, L. RIVERS VINE, and C. F. GORDON CUMMING. Crown 8vo, cloth, 1s. 6d.

Holmes (Oliver Wendell), Works by.
The Autocrat of the Breakfast-Table. Illustrated by J. GORDON THOMSON. Post 8vo, cloth limp, 2s. 6d. Another Edition, post 8vo, cloth, 2s.
The Autocrat of the Breakfast-Table and The Professor at the Breakfast-Table. In One Vol. Post 8vo, half-bound, 2s.

Hooper (Mrs. Geo.).—The House of Raby. Post 8vo, boards, 2s.

Hood's (Thomas) Choice Works in Prose and Verse. With Life of the Author, Portrait, and 200 Illustrations. Crown 8vo, cloth, 3s. 6d.
Hood's Whims and Oddities. With 85 Illustrations. Post 8vo, half-bound, 2s.

Hook's (Theodore) Choice Humorous Works; including his Ludicrous Adventures, Bons Mots, Puns, and Hoaxes. With a Life. A New Edition, with a Frontispiece. Crown 8vo, cloth, 3s. 6d.

Hopkins (Tighe), Novels by.
For Freedom. Crown 8vo, cloth, 6s.

Crown 8vo, cloth, 3s. 6d. each.
'Twixt Love and Duty. With a Frontispiece. | **The Incomplete Adventurer.**
The Nugents of Carriconna. | **Nell Haffenden.** With 8 Illustrations by C. GREGORY.

Horne (R. Hengist).—Orion: An Epic Poem. With Photograph Portrait by SUMMERS. Tenth Edition. Crown 8vo, cloth extra, 7s.

Hornung (E. W.).—The Shadow of the Rope. Crown 8vo, cloth, gilt top, 6s.

Hugo (Victor).—The Outlaw of Iceland (Han d'Islande). Translated by Sir GILBERT CAMPBELL. Crown 8vo, cloth, 3s. 6d.

Hume (Fergus), Novels by.
The Lady from Nowhere. Crown 8vo, cloth, 3s. 6d.; picture cloth, flat back, 2s
The Millionaire Mystery. Crown 8vo, cloth, 3s. 6d.

Hungerford (Mrs., Author of 'Molly Bawn'), Novels by.
Crown 8vo, cloth extra, 3s. 6d. each; post 8vo, illustrated boards, 2s. each; cloth limp, 2s. 6d. each.

A Maiden All Forlorn.	Peter's Wife.	An Unsatisfactory Lover.
In Durance Vile.	Lady Patty.	The Professor's Experiment.
Marvel.	Lady Verner's Flight.	The Three Graces.
A Modern Circe.	The Red-House Mystery.	Nora Creina.
April's Lady.		A Mental Struggle.

Crown 8vo, cloth extra, 3s. 6d. each.
An Anxious Moment. | The Coming of Chloe. | A Point of Conscience. | Lovice.

Hunt's (Leigh) Essays: A Tale for a Chimney Corner, &c. Edited by EDMUND OLLIER. Post 8vo, half-bound, 2s.

Hunt (Mrs. Alfred), Novels by.
Crown 8vo, cloth extra, 3s. 6d. each; post 8vo, illustrated boards, 2s. each.
The Leaden Casket. | **Self-Condemned.** | **That Other Person.**
Mrs. Juliet. Crown 8vo, cloth extra, 3s. 6d.

Hutchison (W. M.).—Hints on Colt-breaking. With 25 Illustrations. Crown 8vo, cloth extra, 3s. 6d.

Hydrophobia: An Account of M. PASTEUR's System; The Technique of his Method, and Statistics. By RENAUD SUZOR, M.B. Crown 8vo, cloth extra, 6s.

Idler Magazine (The). Edited by ROBERT BARR. Profusely Illustrated. 6d. Monthly.

Impressions (The) of Aureole. Post 8vo, cloth, 2s. 6d.

Indoor Paupers. By ONE OF THEM. Crown 8vo, 1s.; cloth, 1s. 6d.

Inman (Herbert) and Hartley Aspden.—The Tear of Kalee. Crown 8vo, cloth, gilt top, 6s.

In Memoriam: Verses for every Day in the Year. Selected and arranged by LUCY RIDLEY. Small square 8vo, cloth, 2s. 6d. net; leather, 3s. 6d. net.

Innkeeper's Handbook (The) and Licensed Victualler's Manual. By J. TREVOR-DAVIES. A New Edition. Crown 8vo, cloth, 2s.

Irish Wit and Humour, Songs of. Collected and Edited by A. PERCEVAL GRAVES. Post 8vo, cloth limp, 2s. 6d.

Irving (Sir Henry): A Record of over Twenty Years at the Lyceum. By PERCY FITZGERALD. With Portrait. Crown 8vo, cloth, 1s. 6d.

James (C. T. C.).—A Romance of the Queen's Hounds. Post 8vo, cloth limp, 1s. 6d.

Jameson (William).—My Dead Self. Post 8vo, cloth, 2s. 6d.

Japp (Alex. H., LL.D.).—Dramatic Pictures, &c. Cr. 8vo, cloth, 5s.

Jennings (Henry J.), Works by.
Curiosities of Criticism. Post 8vo, cloth limp, 2s. 6d.
Lord Tennyson: A Biographical Sketch. With Portrait. Post 8vo, cloth, 1s. 6d.

Jefferies (Richard), Books by.

The Open Air. Post 8vo, cloth, 2s. 6d.

Crown 8vo, buckram. 6s. each ; post 8vo, cloth limp, 2s. 6d. each.

Nature near London. | The Life of the Fields.

Also, the LARGE TYPE, FINE PAPER EDITION of The Life of the Fields. Pott 8vo, cloth, gilt top, 2s. net ; leather, gilt edges, 3s. net.

The Eulogy of Richard Jefferies. By Sir WALTER BESANT. With a Photograph Portrait. Crown 8vo, cloth extra, 6s.

Jerome (Jerome K.), Books by.

Stageland. With 64 Illustrations by J. BERNARD PARTRIDGE. Fcap. 4to, picture cover, 1s.
John Ingerfield, &c. With 9 Illusts. by A. S. BOYD and JOHN GULICH. Fcap. 8vo, pic. cov. 1s. 6d.

Jerrold (Douglas).—The Barber's Chair; and The Hedgehog

Letters. Post 8vo, printed on laid paper and half-bound. 2s.

Jerrold (Tom), Works by. Post 8vo, 1s. ea. ; cloth limp, 1s. 6d. each.

The Garden that Paid the Rent.
Household Horticulture : A Gossip about Flowers. Illustrated.

Jesse (Edward).—Scenes and Occupations of a Country Life.

Post 8vo, cloth limp, 2s.

Jones (William, F.S.A.), Works by. Cr. 8vo, cl. extra, 3s. 6d. each.

Finger-Ring Lore : Historical, Legendary, and Anecdotal. With Hundreds of Illustrations.
Crowns and Coronations : A History of Regalia. With 91 Illustrations.

Jonson's (Ben) Works. With Notes Critical and Explanatory, and

a Biographical Memoir by WILLIAM GIFFORD. Edited by Colonel CUNNINGHAM. Three Vols. crown 8vo, cloth extra, 3s. 6d. each.

Josephus, The Complete Works of. Translated by WHISTON. Con-

taining 'The Antiquities of the Jews' and 'The Wars of the Jews.' With 52 Illustrations and Maps. Two Vols., demy 8vo, half-cloth, 12s. 6d.

Kempt (Robert).—Pencil and Palette: Chapters on Art and Artists.

Post 8vo, cloth limp, 2s. 6d.

Kershaw (Mark). — Colonial Facts and Fictions: Humorous

Sketches. Post 8vo, illustrated boards, 2s. ; cloth, 2s. 6d.

King (R. Ashe), Novels by. Post 8vo, illustrated boards, 2s. each.

'The Wearing of the Green.' | Passion's Slave. | Bell Barry.

A Drawn Game. Crown 8vo, cloth, 3s. 6d. ; post 8vo, illustrated boards, 2s.

Kipling Primer (A). Including Biographical and Critical Chapters,

an Index to Mr. Kipling's principal Writings, and Bibliographies. By F. L. KNOWLES, Editor of 'The Golden Treasury of American Lyrics.' With Two Portraits. Crown 8vo, cloth, 3s. 6d.

Knight (William, M.R.C.S., and Edward, L.R.C.P.). — The

Patient's Vade Mecum : How to Get Most Benefit from Medical Advice. Cr. 8vo, cloth, 1s. 6d.

Knights (The) of the Lion: A Romance of the Thirteenth Century.

Edited, with an Introduction, by the MARQUESS OF LORNE, K.T. Crown 8vo, cloth extra, 6s.

Lambert (George).—The President of Boravia. Crown 8vo, cl., 3s. 6d.

Lamb's (Charles) Complete Works in Prose and Verse, including

'Poetry for Children' and 'Prince Dorus.' Edited, with Notes and Introduction, by R. H. SHEP-HERD. With Two Portraits and Facsimile of the 'Essay on Roast Pig.' Crown 8vo, cloth, 3s. 6d.
The Essays of Elia. Post 8vo, printed on laid paper and half-bound, 2s.
Little Essays : Sketches and Characters by CHARLES LAMB, selected from his Letters by PERCY FITZGERALD. Post 8vo, cloth limp, 2s. 6d.
The Dramatic Essays of Charles Lamb. With Introduction and Notes by BRANDER MAT-THEWS, and Steel-plate Portrait. Fcap. 8vo, half-bound, 2s. 6d.

Landor (Walter Savage).—Citation and Examination of William

Shakspeare, &c. before Sir Thomas Lucy, touching Deer-stealing, 19th September, 1582. To which is added, A Conference of Master Edmund Spenser with the Earl of Essex, touching the State of Ireland, 1595. Fcap. 8vo, half-Roxburghe, 2s. 6d.

Lane (Edward William).—The Thousand and One Nights, com-

monly called in England The Arabian Nights' Entertainments. Translated from the Arabic, with Notes. Illustrated with many hundred Engravings from Designs by HARVEY. Edited by EDWARD STANLEY POOLE. With Preface by STANLEY LANE-POOLE. Three Vols., demy 8vo, cloth, 7s. 6d. ea.

Larwood (Jacob), Works by.

Anecdotes of the Clergy. Post 8vo, laid paper, half cloth, 2s.
Theatrical Anecdotes. Post 8vo, cloth limp, 2s. 6d.
Humour of the Law : Forensic Anecdotes. Post 8vo, cloth, 2s.

Lehmann (R. C.), Works by. Post 8vo, cloth, 1s. 6d. each.

Harry Fludyer at Cambridge.
Conversational Hints for Young Shooters : A Guide to Polite Talk.

Leigh (Henry S.).—Carols of Cockayne. Printed on hand-made paper, bound in buckram, 5s.

Leland (C. Godfrey).—A Manual of Mending and Repairing. With Diagrams. Crown 8vo, cloth, 5s.

Lepelletier (Edmond). — Madame Sans-Gène. Translated from the French by JOHN DE VILLIERS. Post 8vo, cloth, 3s. 6d. ; picture boards, 2s.

Leys (John K.), Novels by.
The Lindsays. Post 8vo, picture bds., 2s. | A Sore Temptation. Cr. 8vo, cloth, gilt top, 6s.

Lilburn (Adam).—A Tragedy in Marble. Crown 8vo, cloth, 3s. 6d.

Lindsay (Harry, Author of 'Methodist Idylls'), Novels by.
Crown 8vo, cloth, 3s. 6d. each.
Rhoda Roberts. | The Jacobite: A Romance of the Conspiracy of 'The Forty.'
Crown 8vo, cloth, gilt top, 6s. each.
Judah Pyecroft, Puritan. | The Story of Leah.

Linton (E. Lynn), Works by.
An Octave of Friends. Crown 8vo, cloth, 3s. 6d.
Crown 8vo, cloth extra, 3s. 6d. each ; post 8vo, illustrated boards, 2s. each.

Patricia Kemball. Ione.	Under which Lord ? With 12 Illustrations.
The Atonement of Leam Dundas.	'My Love!' Sowing the Wind.
The World Well Lost. With 12 Illusts.	Paston Carew, Millionaire and Miser.
The One Too Many.	Dulcie Everton.	With a Silken Thread.

The Rebel of the Family.
Post 8vo, cloth limp, 2s. 6d. each.
Witch Stories. | Ourselves: Essays on Women.
Freeshooting: Extracts from the Works of Mrs. LYNN LINTON.

Lowe (Charles, M.A.).—Our Greatest Living Soldiers. With 8 Portraits. Crown 8vo, cloth, 3s. 6d.

Lucy (Henry W.).—Gideon Fleyce: A Novel. Crown 8vo, cloth extra, 3s. 6d. ; post 8vo, illustrated boards, 2s.

McCarthy (Justin), Works by.
A History of Our Own Times, from the Accession of Queen Victoria to the General Election of 1880. LIBRARY EDITION. Four Vols., demy 8vo, cloth extra, 12s. each.—Also a POPULAR EDITION, in Four Vols., crown 8vo, cloth extra, 6s. each.—And the JUBILEE EDITION, with an Appendix of Events to the end of 1886, in Two Vols., large crown 8vo, cloth extra, 7s. 6d. each.
A History of Our Own Times, Vol. V., from 1880 to the Diamond Jubilee. Demy 8vo, cloth extra, 12s. ; or crown 8vo, cloth, 6s.
A History of Our Own Times, Vol. VI., from the Diamond Jubilee, 1897, to the Accession of King Edward VII. Demy 8vo, cloth, 12s.   [Shortly.
A Short History of Our Own Times. One Vol., crown 8vo, cloth extra, 6s.—Also a CHEAP POPULAR EDITION, post 8vo, cloth limp, 2s. 6d.
A History of the Four Georges and of William the Fourth. By JUSTIN McCARTHY and JUSTIN HUNTLY McCARTHY. Four Vols., demy 8vo, cloth extra, 12s. each.
The Reign of Queen Anne. 2 vols., demy 8vo, cloth, 12s. each.
Reminiscences. With a Portrait. Two Vols., demy 8vo, cloth, 24s.
Crown 8vo, cloth extra, 3s. 6d. each ; post 8vo, illustrated boards, 2s. each ; cloth limp, 2s. 6d. each

The Waterdale Neighbours.	Donna Quixote. With 12 Illustrations.
My Enemy's Daughter.	The Comet of a Season.
A Fair Saxon. Linley Rochford.	Maid of Athens. With 12 Illustrations.
Dear Lady Disdain.	The Dictator.	Camiola: A Girl with a Fortune.
Miss Misanthrope. With 12 Illustrations.	Red Diamonds. The Riddle Ring.

Crown 8vo, cloth, 3s. 6d. each.
The Three Disgraces, and other Stories. | Mononia: A Love Story of 'Forty-eight.'
'The Right Honourable.' By JUSTIN McCARTHY and Mrs. CAMPBELL PRAED. Crown 8vo, cloth extra, 6s.

McCarthy (Justin Huntly), Works by.
The French Revolution. (Constituent Assembly, 1789-91). Four Vols., demy 8vo, cloth, 12s. each.
An Outline of the History of Ireland. Crown 8vo, 1s. ; cloth, 1s. 6d.
Ireland Since the Union: Sketches of Irish History, 1798-1886. Crown 8vo, cloth, 6s.
Hafiz in London: Poems. Small 8vo, gold cloth, 3s. 6d.
Our Sensation Novel. Crown 8vo, picture cover, 1s. ; cloth limp, 1s. 6d.
Doom: An Atlantic Episode. Crown 8vo, picture cover, 1s.
Dolly: A Sketch. Crown 8vo, picture cover, 1s.
Lily Lass: A Romance. Crown 8vo, picture cover, 1s. ; cloth limp, 1s. 6d.
A London Legend. Crown 8vo, cloth, 3s. 6d.

MacColl (Hugh), Novels by.
Mr. Stranger's Sealed Packet. Post 8vo, illustrated boards, 2s.
Ednor Whitlock. Crown 8vo, cloth extra, 6s.

Macdonell (Agnes).—Quaker Cousins. Post 8vo, boards, 2s.

MacGregor (Robert).—Pastimes and Players: Notes on Popular Games. Post 8vo, cloth limp, 2s. 6d.

Machray (Robert), Novels by. Crown 8vo, cloth, gilt top, 6s. each.
A Blow over the Heart. | The Mystery of Lincoln's Inn.

MacDonald (George, LL.D.), Books by.

Works of Fancy and Imagination. Ten Vols., 16mo, cloth, gilt edges, in cloth case, 21s.; or the Volumes may be had separately, in Grolier cloth, at 2s. 6d. each.

Vol. I. WITHIN AND WITHOUT.—THE HIDDEN LIFE.
  " II. THE DISCIPLE.—THE GOSPEL WOMEN.—BOOK OF SONNETS.—ORGAN SONGS.
  " III. VIOLIN SONGS.—SONGS OF THE DAYS AND NIGHTS.—A BOOK OF DREAMS.—ROADSIDE POEMS.—POEMS FOR CHILDREN.
  " IV. PARABLES.—BALLADS.—SCOTCH SONGS.
  " V. & VI. PHANTASTES: A Faerie Romance. | Vol. VII. THE PORTENT.
  " VIII. THE LIGHT PRINCESS.—THE GIANT'S HEART.—SHADOWS.
  " IX. CROSS PURPOSES.—THE GOLDEN KEY.—THE CARASOYN.—LITTLE DAYLIGHT.
  " X. THE CRUEL PAINTER.—THE WOW O' RIVVEN.—THE CASTLE.—THE BROKEN SWORDS.—THE GRAY WOLF.—UNCLE CORNELIUS.

Poetical Works of George MacDonald. Collected and Arranged by the Author. Two Vols. crown 8vo, buckram, 12s.

A Threefold Cord. Edited by GEORGE MACDONALD. Post 8vo, cloth, 5s.

Phantastes: A Faerie Romance. With 25 Illustrations by J. BELL. Crown 8vo, cloth extra, 3s. 6d.
Heather and Snow: A Novel. Crown 8vo, cloth extra, 3s. 6d.; post 8vo, illustrated boards, 2s.
Lilith: A Romance. SECOND EDITION. Crown 8vo, cloth extra, 6s.

Mackay (Charles, LL.D.). — Interludes and Undertones; or,
Music at Twilight. Crown 8vo, cloth extra 6s.

Mackenna (Stephen J.) and J. Augustus O'Shea.—Brave Men
in Action: Thrilling Stories of the British Flag. With 8 Illustrations by STANLEY L. WOOD. Small demy 8vo, cloth, gilt edges, 5s.

Maclise Portrait Gallery (The) of Illustrious Literary Charac-
ters: 85 Portraits by DANIEL MACLISE; with Memoirs—Biographical, Critical, Bibliographical, and Anecdotal—illustrative of the Literature of the former half of the Present Century, by WILLIAM BATES, B.A. Crown 8vo, cloth extra, 7s. 6d.

Macquoid (Mrs.), Works by.   Square 8vo, cloth extra, 6s. each.

In the Ardennes. With 50 Illustrations by THOMAS R. MACQUOID.
Pictures and Legends from Normandy and Brittany. 34 Illusts. by T. R. MACQUOID.
Through Normandy. With 92 Illustrations by T. R. MACQUOID, and a Map.
About Yorkshire. With 67 Illustrations by T. R. MACQUOID.

Magician's Own Book, The: Performances with Eggs, Hats, &c.
Edited by W. H. CREMER. With 200 Illustrations. Crown 8vo, cloth extra, 4s. 6d.

Magic Lantern, The, and its Management: Including full Practical
Directions. By T. C. HEPWORTH. With 10 Illustrations. Crown 8vo, 1s.; cloth, 1s. 6d.

Magna Charta: An Exact Facsimile of the Original in the British
Museum, 3 feet by 2 feet, with Arms and Seals emblazoned in Gold and Colours, 5s.

Mallory (Sir Thomas). — Mort d'Arthur: The Stories of King
Arthur and of the Knights of the Round Table. (A Selection.) Edited by B. MONTGOMERIE RANKING. Post 8vo, cloth limp, 2s.

Mallock (W. H.), Works by.

The New Republic. Post 8vo, cloth, 3s. 6d.; picture boards, 2s.
The New Paul and Virginia: Positivism on an Island. Post 8vo, cloth, 2s. 6d.

Poems. Small 4to, parchment, 8s. | **Is Life Worth Living?** Crown 8vo, cloth extra, 6s.

Margueritte (Paul and Victor).—The Disaster. Translated by
FREDERIC LEES. Crown 8vo, cloth, 3s. 6d.

Marlowe's Works. Including his Translations. Edited, with Notes
and Introductions, by Colonel CUNNINGHAM. Crown 8vo, cloth extra, 3s. 6d.

Mason (Finch).—Annals of the Horse-Shoe Club. With 5 Illus-
trations by the AUTHOR. Crown 8vo, cloth, gilt top, 6s.

Massinger's Plays. From the Text of WILLIAM GIFFORD. Edited
by Col. CUNNINGHAM. Crown 8vo, cloth extra, 3s. 6d.

Matthews (Brander).—A Secret of the Sea, &c. Post 8vo, illus-
trated boards, 2s.; cloth limp, 2s. 6d.

Max O'Rell, Books by. Crown 8vo, cloth, 3s. 6d. each.
Her Royal Highness Woman. | **Between Ourselves.**
Rambles in Womanland.

Meade (L. T.), Novels by.

A Soldier of Fortune. Crown 8vo, cloth, 3s. 6d.; post 8vo, illustrated boards, 2s.

Crown 8vo, cloth, 3s. 6d. each.

The Voice of the Charmer. With 8 Illustrations.		**An Adventuress.**
In an Iron Grip.	**On the Brink of a Chasm.**	**The Blue Diamond.**
The Siren.	**The Way of a Woman.**	**A Stumble by the Way.**
Dr. Rumsey's Patient.	**A Son of Ishmael.**	

Crown 8vo, cloth, gilt top, 6s. each.
This Troublesome World. | **Rosebury.**

Merivale (Herman).—Bar, Stage, and Platform: Autobiographic
Memories. With a Portrait. Crown 8vo, cloth, gilt top, 6s.

Merrick (Leonard), Novels by.
The Man who was Good. Post 8vo, picture boards, 2s.

Crown 8vo, cloth, 3s. 6d. each.
This Stage of Fools. | Cynthia: A Daughter of the Philistines.

Mexican Mustang (On a), through Texas to the Rio Grande. By
A. E. SWEET and J. ARMOY KNOX. With 265 Illustrations. Crown 8vo, cloth extra, 7s. 6d.

Miller (Mrs. F. Fenwick).—Physiology for the Young; or, The
House of Life. With numerous Illustrations. Post 8vo, cloth limp, 2s. 6d.

Milton (J. L.).—The Bath in Diseases of the Skin. Post 8vo,
1s.; cloth, 1s. 6d.

Minto (Wm.).—Was She Good or Bad? Crown 8vo, cloth, 1s. 6d.

Mitchell (Edmund), Novels by.
The Lone Star Rush. With 8 Illustrations by NORMAN H. HARDY. Crown 8vo, cloth, 3s. 6d.

Crown 8vo, cloth, gilt top, 6s. each.
Only a Nigger. | The Belforts of Culben.

Crown 8vo. picture cloth, flat backs, 2s. each.
Plotters of Paris. | The Temple of Death. | Towards the Eternal Snows.

Mitford (Bertram), Novels by. Crown 8vo, cloth extra, 3s. 6d. each.
The Gun-Runner: A Romance of Zululand. With a Frontispiece by STANLEY L. WOOD
Renshaw Fanning's Quest. With a Frontispiece by STANLEY L. WOOD.
The Triumph of Hilary Blachland.

Crown 8vo, cloth, 3s. 6d. each; picture cloth, flat backs, 2s. each.
The Luck of Gerard Ridgeley.
The King's Assegai. With Six full-page Illustrations by STANLEY L. WOOD.

Haviland's Chum. Crown 8vo, cloth, gilt top, 6s.

Molesworth (Mrs.).—Hathercourt Rectory. Crown 8vo, cloth,
3s. 6d.; post 8vo, illustrated boards, 2s.

Moncrieff (W. D. Scott-).—The Abdication: An Historical Drama.
With Seven Etchings by JOHN PETTIE, W. Q. ORCHARDSON, J. MACWHIRTER, COLIN HUNTER,
R. MACBETH and TOM GRAHAM. Imperial 4to, buckram, 21s.

Montagu (Irving).—Things I Have Seen in War. With 16 full-
page Illustrations. Crown 8vo, cloth, 6s.

Moore (Thomas), Works by.
The Epicurean; and Alciphron. Post 8vo, half-bound, 2s.
Prose and Verse; including Suppressed Passages from the MEMOIRS OF LORD BYRON. Edited
by R. H. SHEPHERD. With Portrait. Crown 8vo, cloth extra, 7s. 6d.

Morrow (W. C.).—Bohemian Paris of To-Day. With 106 Illustrations by EDOUARD CUCUEL. Small demy 8vo, cloth, gilt top, 6s.

Muddock (J. E.), Stories by. Crown 8vo, cloth, 3s. 6d. each.
Basile the Jester. With Frontispiece by STANLEY WOOD.
Young Lochinvar. | The Golden Idol.

Post 8vo, illustrated boards, 2s. each.
The Dead Man's Secret. | From the Bosom of the Deep.

Stories Weird and Wonderful. Post 8vo, illustrated boards, 2s.; cloth, 2s. 6d.
Maid Marian and Robin Hood. With 12 Illustrations by S. L. WOOD. Crown 8vo, cloth extra,
3s. 6d.; picture cloth, flat back, 2s.

Murray (D. Christie), Novels by.
Crown 8vo, cloth extra, 3s. 6d. each; post 8vo, illustrated boards, 2s. each.
A Life's Atonement. | A Model Father. | Bob Martin's Little Girl.
Joseph's Coat. 12 Illusts. | Old Blazer's Hero. | Time's Revenges.
Coals of Fire. 3 Illusts. | Cynic Fortune. Frontisp. | A Wasted Crime.
Val Strange. | By the Gate of the Sea. | In Direst Peril.
Hearts. | A Bit of Human Nature. | Mount Despair.
The Way of the World. | First Person Singular. | A Capful o' Nails.
The Making of a Novelist: An Experiment in Autobiography. With a Collotype Portrait. Cr.
8vo, buckram, 3s. 6d.
My Contemporaries in Fiction. Crown 8vo, buckram, 3s. 6d.
His Own Ghost. Crown 8vo, cloth, 3s. 6d.; picture cloth, flat back, 2s.

Crown 8vo, cloth, 3s. 6d. each.
This Little World. | A Race for Millions. | The Church of Humanity
Tales in Prose and Verse. With Frontispiece by ARTHUR HOPKINS.

Crown 8vo, cloth, gilt top, 6s. each.
Despair's Last Journey. | Verona's Father.

Joseph's Coat. POPULAR EDITION, medium 8vo, 6d.

Murray (D. Christie) and Henry Herman, Novels by.
Crown 8vo, cloth extra, 3s. 6d. each; post 8vo, illustrated boards, 2s. each.
One Traveller Returns. | The Bishops' Bible.
Paul Jones's Alias, &c. With Illustrations by A. FORESTIER and G. NICOLET.

Murray (Henry), Novels by.
Post 8vo, cloth, 2s. 6d. each.

A Game of Bluff. | A Song of Sixpence.

Newbolt (H.).—Taken from the Enemy. Post 8vo, leatherette, 1s.

Nisbet (Hume), Books by.
'Bail Up.' Crown 8vo, cloth extra, 3s. 6d.; post 8vo, illustrated boards, 2s.
Dr. Bernard St. Vincent. Post 8vo, illustrated boards, 2s.
Lessons in Art. With 21 Illustrations. Crown 8vo. cloth extra, 2s. 6d.

Norris (W. E.), Novels by. Crown 8vo, cloth, 3s. 6d. each ; post 8vo, picture boards, 2s. each.
Saint Ann's. | Billy Bellew. With a Frontispiece by F. H. TOWNSEND.
Miss Wentworth's Idea. Crown 8vo, cloth, 3s. 6d.

Ohnet (Georges), Novels by. Post 8vo, illustrated boards, 2s. each.
Doctor Rameau. | A Last Love.
A Weird Gift. Crown 8vo, cloth, 3s. 6d.; post 8vo, picture boards, 2s.
Love's Depths. Translated by F. ROTHWELL. Crown 8vo, cloth, 3s. 6d.
The Woman of Mystery. Translated by F. ROTHWELL. Crown 8vo, cloth, gilt top, 6s.

Oliphant (Mrs.), Novels by. Post 8vo, illustrated boards, 2s. each.
The Primrose Path. | Whiteladies. | The Greatest Heiress in England.
The Sorceress. Crown 8vo, cloth, 3s. 6d.

Orrock (James), Painter, Connoisseur, Collector. By BYRON WEBBER. In Two Handsome Volumes, small folio, illustrated with nearly One Hundred Photogravure Plates and a profusion of Drawings reproduced in half-tone. In a handsome binding designed by Sir J. D. LINTON, P.R.I. Price, in buckram gilt, Ten Guineas net. The Edition for sale is strictly limited to Five Hundred Numbered Copies.

O'Shaughnessy (Arthur), Poems by :
Fcap. 8vo, cloth extra, 7s. 6d. each.

Music and Moonlight. | Songs of a Worker.
Lays of France. Crown 8vo, cloth extra, 10s. 6d.

Ouida, Novels by. Cr. 8vo, cl., 3s. 6d. ea.; post 8vo, illust. bds., 2s. ea.

Held in Bondage.	A Dog of Flanders.	In Maremma.	Wanda.
Tricotrin.	Pascarel.	Signa.	Bimbi.
Strathmore.	Chandos.	Two Wooden Shoes.	Frescoes.	Othmar.
Cecil Castlemaine's Gage	In a Winter City.	Princess Napraxine.
Under Two Flags.	Ariadne.	Friendship.	Guilderoy.	Ruffino.
Puck.	Idalia.	A Village Commune.	Two Offenders.
Folle-Farine.	Moths.	Pipistrello.	Santa Barbara.

POPULAR EDITIONS, medium 8vo, 6d. each.
Under Two Flags. | Moths. | Held in Bondage. | Puck. | Strathmore.

Syrlin. Crown 8vo, cloth, 3s. 6d.; post 8vo, picture cloth, flat back, 2s.; illustrated boards, 2s.
The Waters of Edera. Crown 8vo, cloth, 3s. 6d.; picture cloth, flat back, 2s.
Wisdom, Wit, and Pathos, selected from the Works of OUIDA by F. SYDNEY MORRIS. Post 8vo, cloth extra, 5s.—CHEAP EDITION, illustrated boards, 2s.

Pain (Barry).—Eliza's Husband. Fcap. 8vo, picture cover, 1s.

Palmer (W. T.), Books by. Crown 8vo, cloth, gilt top, 6s. each.
Lake Country Rambles. With a Frontispiece.
In Lakeland Dells and Fells. With a Frontispiece.

Payn (James), Novels by.
Crown 8vo, cloth extra, 3s. 6d. each ; post 8vo, illustrated boards, 2s. each.

Lost Sir Massingberd.	The Family Scapegrace.
A County Family.	Holiday Tasks.
Less Black than We're Painted.	The Talk of the Town. With 12 Illusts.
By Proxy.	For Cash Only.	The Mystery of Mirbridge.
High Spirits.	The Word and the Will.
A Confidential Agent. With 12 Illusts.	The Burnt Million.
A Grape from a Thorn. With 12 Illusts.	Sunny Stories.	A Trying Patient.

Post 8vo, illustrated boards, 2s. each.

Humorous Stories.	From Exile.	Found Dead.	Gwendoline's Harvest
The Foster Brothers.	Mirk Abbey.	A Marine Residence.
Married Beneath Him.	The Canon's Ward.
Bentinck's Tutor.	Walter's Word.	Not Wooed, But Won.
A Perfect Treasure.	Two Hundred Pounds Reward.
Like Father, Like Son.	The Best of Husbands.
A Woman's Vengeance.	Halves.	What He Cost Her.
Carlyon's Year.	Cecil's Tryst.	Fallen Fortunes.	Kit: A Memory.
Murphy's Master.	At Her Mercy.	Under One Roof.	Glow-worm Tales.
The Clyffards of Clyffe.	A Prince of the Blood.
Some Private Views.	

A Modern Dick Whittington; or, A Patron of Letters. With a Portrait of the Author. Crown 8vo, cloth, 3s. 6d.; picture cloth, flat back, 2s.
In Peril and Privation. With 17 Illustrations. Crown 8vo, cloth, 3s. 6d.
Notes from the ' News.' Crown 8vo, cloth, 1s. 6d.

Payne (Will).—Jerry the Dreamer. Crown 8vo, cloth, 3s. 6d.

Pandurang Hari; or, Memoirs of a Hindoo. With Preface by Sir BARTLE FRERE. Post 8vo, illustrated boards, 2s.

Paris Salon, The Illustrated Catalogue of the, for 1902. (Twenty-fourth Year.) With over 300 Illustrations. Demy 8vo, 3s.

Pascal's Provincial Letters. A New Translation, with Historical Introduction and Notes by T. M'CRIE, D.D. Post 8vo, half-cloth, 2s.

Paston Letters (The), 1422-1509. Containing upwards of 600 more Letters than appeared in the original 5-volume issue in 1787-1823. Edited, with Introduction and Notes, by JAMES GAIRDNER, of the Public Record Office. A NEW EDITION. In 6 Volumes, square demy 8vo, art linen, gilt top, 12s. 6d. net per volume (sold only in sets). The Edition will consist of 650 copies, of which only 600 are for sale. The First Volume will be ready in November.

Paul (Margaret A.).—Gentle and Simple. Crown 8vo, cloth, with Frontispiece by HELEN PATERSON, 3s. 6d.; post 8vo, illustrated boards, 2s.

Pennell-Elmhirst (Captain E.).—The Best of the Fun. With 8 Coloured Illustrations by G. D. GILES, and 48 in Black and White by J. STURGESS and G. D. GILES. Medium 8vo, cloth, gilt top, 16s.

Pennell (H. Cholmondeley), Works by. Post 8vo, cloth, 2s. 6d. ea.
Puck on Pegasus. With Illustrations.
Pegasus Re-Saddled. With Ten full-page Illustrations by G. DU MAURIER.
The Muses of Mayfair: Vers de Société. Selected by H. C. PENNELL.

Phelps (E. Stuart), Works by. Post 8vo, cloth, 1s. 6d. each.
An Old Maid's Paradise. | Burglars in Paradise.
Beyond the Gates. Post 8vo, picture cover, 1s. ; cloth, 1s. 6d.
Jack the Fisherman. Illustrated by C. W. REED. Crown 8vo, cloth, 1s. 6d.

Phil May's Sketch-Book. Containing 54 Humorous Cartoons. Crown folio, cloth, 2s. 6d.

Phipson (Dr. T. L.), Books by. Crown 8vo, canvas, gilt top, 5s. each.
Famous Violinists and Fine Violins. | The Confessions of a Violinist.
Voice and Violin: Sketches, Anecdotes, and Reminiscences.

Pilkington (Lionel L.).—Mallender's Mistake. Crown 8vo, cloth, gilt top, 6s.

Planche (J. R.), Works by.
The Pursuivant of Arms. With Six Plates and 209 Illustrations. Crown 8vo, cloth, 7s. 6d.
Songs and Poems, 1819-1879. With Introduction by Mrs. MACKARNESS. Crown 8vo, cloth, 6s.

Plutarch's Lives of Illustrious Men. With Notes and a Life of Plutarch by JOHN and WM. LANGHORNE, and Portraits. Two Vols., demy 8vo, half-cloth 10s. 6d.

Poe's (Edgar Allan) Choice Works: Poems, Stories, Essays. With an Introduction by CHARLES BAUDELAIRE. Crown 8vo, cloth, 3s. 6d.

Pollock (W. H.).—The Charm, and other Drawing-room Plays. By Sir WALTER BESANT and WALTER H. POLLOCK. With 50 Illustrations. Crown 8vo, cloth gilt, 6s.

Pope's Poetical Works. Post 8vo, cloth limp, 2s.

Porter (John).—Kingsclere. Edited by BYRON WEBBER. With 19 full-page and many smaller Illustrations. Cheaper Edition. Demy 8vo, cloth, 7s. 6d.

Praed (Mrs. Campbell), Novels by. Post 8vo, illust. bds., 2s. each.
The Romance of a Station. | The Soul of Countess Adrian.

Crown 8vo, cloth, 3s. 6d. each ; post 8vo, boards, 2s. each.
Outlaw and Lawmaker. | Christina Chard. With Frontispiece by W. PAGET.
Mrs. Tregaskiss. With 8 Illustrations by ROBERT SAUBER.

Crown 8vo, cloth, 3s. 6d. each.
Nulma. | Madame Izan. | 'As a Watch in the Night.'

Price (E. C.).—Valentina. Crown 8vo, cloth, 3s. 6d.

Princess Olga.—Radna: A Novel. Crown 8vo, cloth extra, 6s.

Pryce (Richard).—Miss Maxwell's Affections. Crown 8vo, cloth, with Frontispiece by HAL LUDLOW, 3s. 6d.; post 8vo, illustrated boards, 2s.

Proctor (Richard A.), Works by.
Flowers of the Sky. With 55 Illustrations. Small crown 8vo, cloth extra, 3s. 6d.
Easy Star Lessons. With Star Maps for every Night in the Year. Crown 8vo, cloth, 6s.
Familiar Science Studies. Crown 8vo, cloth extra, 6s.
Saturn and its System. With 13 Steel Plates. Demy 8vo, cloth extra, 10s. 6d.
Mysteries of Time and Space. With numerous Illustrations. Crown 8vo, cloth extra, 6s.
The Universe of Suns, &c. With numerous Illustrations. Crown 8vo, cloth extra, 6s.
Wages and Wants of Science Workers. Crown 8vo, 1s. 6d.

Rambosson (J.).—Popular Astronomy. Translated by C. B. PITMAN. With 10 Coloured Plates and 63 Woodcut Illustrations. Crown 8vo, cloth, 3s. 6d.

Randolph (Col. G.).—Aunt Abigail Dykes. Crown 8vo, cloth, 7s. 6d.

Richardson (Frank), Novels by.
The Man who Lost his Past. With 50 Illusts. by TOM BROWNE, R.I. Cr. 8vo, cloth, 3s. 6d.
Crown 8vo, cloth, gilt top, 6s. each.
The King's Counsel. | Semi-Society.

Riddell (Mrs. J. H.), Novels by.
A Rich Man's Daughter. Crown 8vo, cloth, 3s. 6d.
Weird Stories. Crown 8vo, cloth extra, 3s. 6d. ; post 8vo, illustrated boards, 2s.
Post 8vo, illustrated boards, 2s. each.
The Uninhabited House. | Fairy Water.
The Prince of Wales's Garden Party. | Her Mother's Darling.
The Mystery in Palace Gardens. | The Nun's Curse. | Idle Tales.

Reade's (Charles) Novels.
The New Collected LIBRARY EDITION, complete in Seventeen Volumes, set in new long primer type, printed on laid paper, and elegantly bound in cloth, price 3s. 6d. each.
1. Peg Woffington; and Christie Johnstone.
2. Hard Cash.
3. The Cloister and the Hearth. With a Preface by Sir WALTER BESANT.
4. 'It is Never Too Late to Mend.'
5. The Course of True Love Never Did Run Smooth; and Singleheart and Doubleface.
6. The Autobiography of a Thief; Jack of all Trades; A Hero and a Martyr; and The Wandering Heir.
7. Love Me Little, Love me Long.
8. The Double Marriage.
9. Griffith Gaunt.
10. Foul Play.
11. Put Yourself in His Place.
12. A Terrible Temptation.
13. A Simpleton.
14. A Woman-Hater.
15. The Jilt, and other Stories; and Good Stories of Man and other Animals.
16. A Perilous Secret.
17. Readiana; and Bible Characters.

In Twenty-one Volumes, post 8vo, illustrated boards, 2s. each.
Peg Woffington. | Christie Johnstone.
'It is Never Too Late to Mend.'
The Course of True Love Never Did Run Smooth.
The Autobiography of a Thief; Jack of all Trades; and James Lambert.
Love Me Little, Love Me Long.
The Double Marriage.
The Cloister and the Hearth.
Hard Cash. | Griffith Gaunt.
Foul Play. | Put Yourself in His Place.
A Terrible Temptation.
A Simpleton. | The Wandering Heir.
A Woman-Hater.
Singleheart and Doubleface.
Good Stories of Man and other Animals.
The Jilt, and other Stories.
A Perilous Secret. | Readiana.

LARGE TYPE, FINE PAPER EDITIONS. Pott 8vo, cl., gilt top, 2s. net ea. ; leather, gilt edges, 3s. net ea.
The Cloister and the Hearth. | 'It is Never Too Late to Mend.'

POPULAR EDITIONS, medium 8vo, 6d. each.
'It is Never Too Late to Mend.' | The Cloister and the Hearth. | Foul Play.
Peg Woffington; and Christie Johnstone. | Hard Cash. | Griffith Gaunt.

Christie Johnstone. With Frontispiece. Choicely printed in Elzevir style. Fcap. 8vo, half-Roxb. 2s. 6d.
Peg Woffington. Choicely printed in Elzevir style. Fcap. 8vo, half-Roxburghe, 2s. 6d.
The Cloister and the Hearth. NEW ILLUSTRATED EDITION, with 16 Photogravure and 84 half-tone Illustrations by MATT B. HEWERDINE. Small 4to, cloth gilt and gilt top, 10s. 6d. net.— Also in Four Vols., post 8vo, with an Introduction by Sir WALTER BESANT, and a Frontispiece to each Vol., buckram, gilt top, 6s. the set.
Bible Characters. Fcap. 8vo, leatherette, 1s.
Selections from the Works of Charles Reade. With an Introduction by Mrs. ALEX. IRELAND. Post 8vo, cloth limp, 2s. 6d.

Rimmer (Alfred), Works by. Large crown 8vo, cloth, 3s. 6d. each.
Rambles Round Eton and Harrow. With 52 Illustrations by the Author.
About England with Dickens. With 58 Illustrations by C. A. VANDERHOOF and A. RIMMER.

Rives (Amelie), Stories by. Crown 8vo, cloth, 3s. 6d. each.
Barbara Dering. | Meriel: A Love Story.

Robinson Crusoe. By DANIEL DEFOE. With 37 Illustrations by GEORGE CRUIKSHANK. Post 8vo, half-cloth, 2s.

Robinson (F. W.), Novels by.
Women are Strange. Post 8vo, illustrated boards, 2s.
The Hands of Justice. Crown 8vo, cloth extra, 3s. 6d. ; post 8vo illustrated boards, 2s.
The Woman in the Dark. Crown 8vo, cloth, 3s. 6d. ; post 8vo, illustrated boards, 2s.

Robinson (Phil), Works by. Crown 8vo, cloth extra, 6s. each.
The Poets' Birds. | The Poets' Beasts. | The Poets' Reptiles, Fishes, and Insects.

Roll of Battle Abbey, The: A List of the Principal Warriors who came from Normandy with William the Conqueror, 1066. Printed in Gold and Colours, 5s.

Rosengarten (A.).—A Handbook of Architectural Styles. Translated by W. COLLETT-SANDARS. With 639 Illustrations. Crown 8vo, cloth extra, 7s. 6d.

Ross (Albert).—A Sugar Princess. Crown 8vo, cloth, 3s. 6d.

Rowley (Hon. Hugh). Post 8vo, cloth, 2s. 6d. each.
Puniana: or, Thoughts Wise and Other-wise : a Collection of the Best Riddles, Conundrums, Jokes, Sells, &c., with numerous Illustrations by the Author.
More Puniana: A Second Collection of Riddles, Jokes, &c. With numerous Illustrations.

Runciman (James), Stories by.
Schools and Scholars. Post 8vo, cloth, 2s. 6d.
Skippers and Shellbacks. Crown 8vo, cloth, 3s. 6d.

Russell (Dora), Novels by.
A Country Sweetheart. Post 8vo, picture boards, 2s. ; picture cloth, flat back, 2s.
The Drift of Fate. Crown 8vo, cloth, 3s. 6d.; picture cloth, flat back, 2s.

Russell (Herbert).—True Blue; or, 'The Lass that Loved a Sailor.'
Crown 8vo, cloth, 3s. 6d.

Russell (Rev. John) and his Out-of-door Life. By E. W. L.
DAVIES. A New Edition, with Illustrations coloured by hand. Royal 8vo, cloth, 16s. net.

Russell (W. Clark), Novels, &c., by.
Crown 8vo, cloth extra, 3s. 6d. each ; post 8vo, illustrated boards, 2s. each ; cloth limp, 2s. 6d. each.

Round the Galley-Fire.	An Ocean Tragedy.
In the Middle Watch.	My Shipmate Louise.
On the Fo'k'sle Head.	Alone on a Wide Wide Sea.
A Voyage to the Cape.	The Good Ship 'Mohock.'
A Book for the Hammock.	The Phantom Death.
The Mystery of the 'Ocean Star.'	Is He the Man? | The Convict Ship.
The Romance of Jenny Harlowe.	Heart of Oak. | The Last Entry.

The Tale of the Ten.
Crown 8vo, cloth, 3s. 6d. each.

A Tale of Two Tunnels.	The Death Ship.

The Ship: Her Story. With 50 Illustrations by H. C. SEPPINGS WRIGHT. Small 4to, cloth, 6s.
The 'Pretty Polly': A Voyage of Incident. With 12 Illustrations by G. E. ROBERTSON.
Large crown 8vo, cloth, gilt edges, 5s.
Overdue. Crown 8vo, cloth, gilt top, 6s.

Saint Aubyn (Alan), Novels by.
Crown 8vo, cloth extra, 3s. 6d. each ; post 8vo, illustrated boards, 2s. each.
A Fellow of Trinity. With a Note by OLIVER WENDELL HOLMES and a Frontispiece.

The Junior Dean.	The Master of St. Benedict's.	To His Own Master.
Orchard Damerel.	In the Face of the World.	The Tremlett Diamonds.

Crown 8vo, cloth, 3s. 6d. each.

The Wooing of May.	A Tragic Honeymoon.	A Proctor's Wooing.
Fortune's Gate.	Gallantry Bower.	Bonnie Maggie Lauder.
Mary Unwin. With 8 Illustrations by PERCY TARRANT.		Mrs. Dunbar's Secret.

Saint John (Bayle).—A Levantine Family. Cr. 8vo, cloth, 3s. 6d.

Sala (George A.).—Gaslight and Daylight. Post 8vo, boards, 2s.

Scotland Yard, Past and Present: Experiences of Thirty-seven Years.
By Ex-Chief-Inspector CAVANAGH. Post 8vo, illustrated boards, 2s.: cloth, 2s. 6d.

Secret Out, The: One Thousand Tricks with Cards; with Entertaining Experiments in Drawing-room or 'White' Magic. By W. H. CREMER. With 300 Illustrations. Crown 8vo, cloth extra, 4s. 6d.

Seguin (L. G.), Works by.
The Country of the Passion Play (Oberammergau) and the Highlands of Bavaria. With Map and 37 Illustrations. Crown 8vo, cloth extra, 3s. 6d.
Walks in Algiers. With Two Maps and 16 Illustrations. Crown 8vo, cloth extra, 6s.

Senior (Wm.).—By Stream and Sea. Post 8vo, cloth, 2s. 6d.

Sergeant (Adeline), Novels by. Crown 8vo, cloth, 3s. 6d. each.

Under False Pretences.	Dr. Endicott's Experiment.

Seymour (Cyril).—The Magic of To-Morrow. Cr. 8vo, cloth, 6s.

Shakespeare for Children: Lamb's Tales from Shakespeare.
With Illustrations, coloured and plain, by J. MOYR SMITH. Crown 4to, cloth gilt, 3s. 6d.

Shakespeare the Boy. With Sketches of the Home and School Life,
the Games and Sports, the Manners, Customs, and Folk-lore of the Time. By WILLIAM J. ROLFE,
Litt.D. A New Edition, with 42 Illustrations, and an INDEX OF PLAYS AND PASSAGES RE-
FERRED TO. Crown 8vo, cloth gilt, 3s. 6d.

Shelley's (Percy Bysshe) Complete Works in Verse and Prose.
Edited, Prefaced, and Annotated by R. HERNE SHEPHERD. Five Vols., crown 8vo, cloth, 3s. 6d. each.
Poetical Works, in Three Vols.:
Vol. I. Introduction by the Editor; Posthumous Fragments of Margaret Nicholson; Shelley's Corre-
spondence with Stockdale; The Wandering Jew; Queen Mab, with the Notes; Alastor,
and other Poems; Rosalind and Helen; Prometheus Unbound; Adonais, &c.
„ II. Laon and Cythna; The Cenci; Julian and Maddalo; Swellfoot the Tyrant; The Witch of
Atlas; Epipsychidion; Hellas.
„ III. Posthumous Poems; The Masque of Anarchy; and other Pieces.
Prose Works, in Two Vols.:
Vol. I. The Two Romances of Zastrozzi and St. Irvyne; the Dublin and Marlow Pamphlets; A Refu-
tation of Deism; Letters to Leigh Hunt, and some Minor Writings and Fragments.
II. The Essays; Letters from Abroad; Translations and Fragments, edited by Mrs. SHELLEY.
With a Biography of Shelley, and an Index of the Prose Works.

Sharp (William).—Children of To-morrow. Crown 8vo, cloth, 6s.

Sherard (R. H.).—Rogues: A Novel.   Crown 8vo, cloth, 1s. 6d.

Sheridan's (Richard Brinsley) Complete Works, with Life and
Anecdotes. Including Drama, Prose and Poetry, Translations, Speeches, Jokes. Cr. 8vo, cloth, 3s. 6d.
The Rivals, The School for Scandal, and other Plays.   Post 8vo, half-bound, 2s.
Sheridan's Comedies: The Rivals and The School for Scandal. Edited, with an Introduction and Notes to each Play, and a Biographical Sketch, by BRANDER MATTHEWS. With
Illustrations.   Demy 8vo, buckram, gilt top, 12s. 6d.

Shiel (M. P.), Novels by.   Crown 8vo, cloth, gilt top, 6s. each.
The Purple Cloud. | Unto the Third Generation.

Sidney's (Sir Philip) Complete Poetical Works, including all
those in 'Arcadia.' With Portrait, Memorial-Introduction, Notes, &c., by the Rev. A. B. GROSART,
D.D.   Three Vols., crown 8vo, cloth boards, 3s. 6d. each.

Signboards: Their History, including Anecdotes of Famous Taverns and
Remarkable Characters. By JACOB LARWOOD and JOHN CAMDEN HOTTEN.   With Coloured Frontispiece and 94 Illustrations.   Crown 8vo, cloth extra, 3s. 6d.

Sims (George R.), Works by.
Post 8vo, illustrated boards, 2s. each; cloth limp, 2s. 6d. each.
The Ring o' Bells. | My Two Wives. | Memoirs of a Landlady.
Tinkletop's Crime. | Tales of To-day. | Scenes from the Show.
Zeph: A Circus Story, &c. | The Ten Commandments: Stories.
Dramas of Life. With 60 Illustrations.

Crown 8vo, picture cover, 1s. each; cloth, 1s. 6d. each.
The Dagonet Reciter and Reader: Being Readings and Recitations in Prose and Verse
selected from his own Works by GEORGE R. SIMS.
The Case of George Candlemas. | Dagonet Ditties. (From The Referee.)
How the Poor Live; and Horrible London. With a Frontispiece by F. BARNARD.
Crown 8vo, leatherette, 1s. | Dagonet Dramas of the Day.   Crown 8vo, 1s.

Crown 8vo, cloth, 3s. 6d. each; post 8vo, picture boards, 2s. each; cloth limp, 2s. 6d. each.
Mary Jane's Memoirs. | Mary Jane Married. | Rogues and Vagabonds.
Dagonet Abroad.

Crown 8vo, cloth, 3s. 6d. each.
Once upon a Christmas Time.   With 8 Illustrations by CHARLES GREEN, R.I.
In London's Heart: A Story of To-day.—Also in picture cloth, flat back, 2s | A Blind Marriage
Without the Limelight: Theatrical Life as it is. | The Small-part Lady, &c.
Biographs of Babylon: Life Pictures of London's Moving Scenes.

Sinclair (Upton).—Prince Hagen: A Phantasy.   Cr. 8vo, cl., 3s. 6d.

Sister Dora: A Biography.   By MARGARET LONSDALE.   With Four
Illustrations.   Demy 8vo, picture cover, 4d.; cloth, 6d.

Sketchley (Arthur).—A Match in the Dark.   Post 8vo, boards, 2s.

Slang Dictionary (The): Etymological, Historical, and Anecdotal.
Crown 8vo, cloth extra, 6s. 6d.

Smart (Hawley), Novels by.
Crown 8vo, cloth 3s. 6d. each; post 8vo, picture boards, 2s. each.
Beatrice and Benedick. | Long Odds.
Without Love or Licence. | The Master of Rathkelly.

Crown 8vo, cloth, 3s. 6d. each.
The Outsider | A Racing Rubber.
The Plunger. Post 8vo, picture boards, 2s.

Smith (J. Moyr), Works by.
The Prince of Argolis. With 130 Illustrations.   Post 8vo, cloth extra, 3s. 6d.
The Wooing of the Water Witch. With numerous Illustrations.   Post 8vo, cloth, 6s.

Snazelleparilla.   Decanted by G. S. EDWARDS.   With Portrait of
G. H. SNAZELLE, and 65 Illustrations by C. LYALL.   Crown 8vo, cloth, 3s. 6d.

Society in London.   Crown 8vo, 1s.; cloth, 1s. 6d.

Speight (T. W.), Novels by.
Post 8vo, illustrated boards, 2s. each.
The Mysteries of Heron Dyke. | The Loudwater Tragedy.
By Devious Ways, &c. | Burgo's Romance.
Hoodwinked; & Sandycroft Mystery. | Quittance in Full.
The Golden Hoop. | Back to Life. | A Husband from the Sea.

Post 8vo, cloth limp, 1s. 6d. each.
A Barren Title. | Wife or No Wife?

Crown 8vo, cloth extra, 3s. 6d. each.
A Secret of the Sea. | The Grey Monk. | The Master of Trenance.
A Minion of the Moon: A Romance of the King's Highway. | Her Ladyship.
The Secret of Wyvern Towers. | The Doom of Siva. | The Web of Fate.
The Strange Experiences of Mr. Verschoyle. | As It was Written.
Stepping Blindfold. Crown 8vo, cloth, gilt top, 6s.

Spalding (T. A., LL.B.).—Elizabethan Demonology: An Essay
on the Belief in the Existence of Devils.   Crown 8vo, cloth extra, 5s.

Somerset (Lord Henry).—Songs of Adieu. Small 4to, Jap. vel., 6s.

Spenser for Children. By M. H. TOWRY. With Coloured Illustrations by WALTER J. MORGAN. Crown 4to, cloth extra, 3s. 6d.

Sprigge (S. Squire).—An Industrious Chevalier. Crown 8vo, cloth, gilt top, 6s.

Spettigue (H. H.).—The Heritage of Eve. Crown 8vo, cloth, 6s.

Stafford (John), Novels by.
Doris and I. Crown 8vo, cloth, 3s. 6d. | Carlton Priors. Crown 8vo, cloth, gilt top, 6s.

Starry Heavens (The): POETICAL BIRTHDAY BOOK. Roy. 16mo, cl., 2s. 6d.

Stag-Hunting with the 'Devon and Somerset:' Chase of the Wild Red Deer on Exmoor. By PHILIP EVERRD. With 70 Illustrations. Crown 4to, cloth, 16s. net.

Stedman (E. C.).—Victorian Poets. Crown 8vo, cloth extra, 9s.

Stephens (Riccardo, M.B.).—The Cruciform Mark: The Strange Story of RICHARD TREGENNA, Bachelor of Medicine (Univ. Edinb.) Crown 8vo, cloth, 3s. 6d.

Stephens (Robert Neilson).—Philip Winwood: A Sketch of the Domestic History of an American Captain in the War of Independence. Crown 8vo, cloth, 3s. 6d.

Sterndale (R. Armitage).—The Afghan Knife: A Novel. Post 8vo, cloth, 3s. 6d.; illustrated boards, 2s.

Stevenson (R. Louis), Works by.
Crown 8vo, buckram, gilt top, 6s. each.
Travels with a Donkey. With a Frontispiece by WALTER CRANE.
An Inland Voyage. With a Frontispiece by WALTER CRANE.
Familiar Studies of Men and Books.
The Silverado Squatters. With Frontispiece by J. D. STRONG.
The Merry Men. | Underwoods: Poems. | Memories and Portraits.
Virginibus Puerisque, and other Papers. | Ballads. | Prince Otto.
Across the Plains, with other Memories and Essays.
Weir of Hermiston. | In the South Seas.
Songs of Travel. Crown 8vo, buckram, 5s.
New Arabian Nights. Crown 8vo, buckram, gilt top, 6s.; post 8vo, illustrated boards, 2s.
—POPULAR EDITION, medium 8vo, 6d.
The Suicide Club; and The Rajah's Diamond. (From NEW ARABIAN NIGHTS.) With Eight Illustrations by W. J. HENNESSY. Crown 8vo, cloth, 3s. 6d.
The Stevenson Reader: Selections from the Writings of ROBERT LOUIS STEVENSON. Edited by LLOYD OSBOURNE. Post 8vo, cloth, 2s. 6d.; buckram, gilt top, 3s. 6d.
The Pocket R.L.S.: Favourite Passages. Small 16mo, cloth, 2s. net; leather, 3s. net.
LARGE TYPE, FINE PAPER EDITIONS. Pott 8vo, cl., gilt top, 2s. net each; leather, gilt edges, 3s. net each.
Virginibus Puerisque, and other Papers. | New Arabian Nights.
Familiar Studies of Men and Books. |
R. L. Stevenson: A Study. By H. B. BAILDON. With 2 Portraits. Crown 8vo, buckram, 6s.

Stockton (Frank R.).—The Young Master of Hyson Hall. With 36 Illustrations by V. H. DAVISSON and C. H. STEPHENS. Crown 8vo, cloth, 3s. 6d.; picture cloth, 2s.

Stories from Foreign Novelists. Crown 8vo, cloth, 3s. 6d.

Strange Manuscript (A) Found in a Copper Cylinder. By JAMES DE MILLE. Crown 8vo, cloth, with 19 Illusts. by GILBERT GAUL, 3s. 6d.; post 8vo, boards, 2s.

Strange Secrets. Told by PERCY FITZGERALD, CONAN DOYLE, FLORENCE MARRYAT, &c. Post 8vo, illustrated boards, 2s.

Strutt (Joseph). — The Sports and Pastimes of the People of England. Edited by WILLIAM HONE. With 140 Illustrations. Crown 8vo, cloth, 3s. 6d.

Sundowner, Stories by.
Told by the Taffrail. Cr. 8vo, cl., 3s. 6d. | The Tale of the Serpent. Cr. 8vo, cl., flat back, 2s.

Surtees (Robert).—Handley Cross; or, Mr. Jorrocks's Hunt. With 79 Illustrations by JOHN LEECH. Post 8vo, cloth, 2s.

Sutro (Alfred).—The Foolish Virgins. Fcap. 8vo, picture cover, 1s.

Swift's (Dean) Choice Works, in Prose and Verse. With Memoir, Portrait, and Facsimiles of the Maps in 'Gulliver's Travels.' Crown 8vo, cloth, 3s. 6d.
Gulliver's Travels, and A Tale of a Tub. Post 8vo, half-bound, 2s.
Jonathan Swift: A Study. By J. CHURTON COLLINS. Crown 8vo, cloth extra, 8s.

Taine's History of English Literature. Translated by HENRY VAN LAUN. Four Vols., demy 8vo, cloth, 30s.—POPULAR EDITION, Two Vols., crown 8vo, cloth, 15s.

Taylor (Bayard).—Diversions of the Echo Club. Post 8vo, cl., 2s.

Swinburne's (Algernon Charles) Works.

Selections from the Poetical Works of A. C. Swinburne. Fcap. 8vo 6s.
Atalanta in Calydon. Crown 8vo, 6s.
Chastelard: A Tragedy. Crown 8vo, 7s.
Poems and Ballads. FIRST SERIES. Cr.8vo,9s.
Poems and Ballads. SECOND SER. Cr.8vo,9s.
Poems & Ballads. THIRD SERIES. Cr.8vo, 7s.
Songs before Sunrise. Crown 8vo, 10s. 6d.
Bothwell: A Tragedy. Crown 8vo, 12s. 6d.
Songs of Two Nations. Crown 8vo, 6s.
George Chapman. (See Vol. II. of G. CHAPMAN'S Works.) Crown 8vo, 3s. 6d.
Essays and Studies. Crown 8vo, 12s.
Erechtheus: A Tragedy. Crown 8vo, 6s.
A Note on Charlotte Bronte. Cr. 8vo, 6s.
A Study of Shakespeare. Crown 8vo, 8s.
Songs of the Springtides. Crown 8vo, 6s

Studies in Song. Crown 8vo, 7s.
Mary Stuart: A Tragedy. Crown 8vo, 8s.
Tristram of Lyonesse. Crown 8vo, 9s.
A Century of Roundels. Small 4to, 8s.
A Midsummer Holiday. Crown 8vo, 7s.
Marino Faliero: A Tragedy. Crown 8vo, 6s.
A Study of Victor Hugo. Crown 8vo, 6s.
Miscellanies. Crown 8vo, 12s.
Locrine: A Tragedy. Crown 8vo, 6s.
A Study of Ben Jonson. Crown 8vo, 7s.
The Sisters: A Tragedy. Crown 8vo, 6s.
Astrophel, &c. Crown 8vo, 7s.
Studies in Prose and Poetry. Cr.8vo, 9s.
The Tale of Balen. Crown 8vo, 7s.
Rosamund, Queen of the Lombards: A Tragedy. Crown 8vo, 6s.
A New Volume of Poems. Cr. 8vo. [Shortly.

Taylor (Tom).—Historical Dramas: 'JEANNE DARC,' ''TWIXT AXE AND CROWN,' 'THE FOOL'S REVENGE,' 'ARKWRIGHT'S WIFE,' 'ANNE BOLEYNE,' 'PLOT AND PASSION.' Crown 8vo, 1s. each.

Temple (Sir Richard, G.C.S.I.).—A Bird's-eye View of Picturesque India. With 32 Illustrations by the Author. Crown 8vo, cloth, gilt top, 6s.

Thackerayana: Notes and Anecdotes. With Coloured Frontispiece and Hundreds of Sketches by WILLIAM MAKEPEACE THACKERAY. Crown 8vo, cloth extra, 3s. 6d.

Thames, A New Pictorial History of the. By A. S. KRAUSSE. With 340 Illustrations. Post 8vo, cloth, 1s. 6d.

Thomas (Annie), Novels by.

The Siren's Web: A Romance of London Society. Crown 8vo, cloth, 3s. 6d.
Comrades True. Crown 8vo, cloth, gilt top, 6s.

Thomas (Bertha), Novels by. Crown 8vo, cloth, 3s. 6d. each.

The Violin-Player. | In a Cathedral City.

Crown 8vo, cloth, gilt top, 6s. each.

The House on the Scar: a Tale of South Devon. | The Son of the House.

Thomson's Seasons, and The Castle of Indolence. With Introduction by ALLAN CUNNINGHAM, and 48 Illustrations. Post 8vo, half-bound, 2s.

Thoreau: His Life and Aims. By H. A. PAGE. With a Portrait and View. Post 8vo buckram, 3s. 6d.

Tompkins (H. W.).—Marsh-Country Rambles. With a Frontispiece. Crown 8vo, cloth, gilt top, 6s.   [Shortly.

Thornbury (Walter), Books by.

The Life and Correspondence of J. M. W. Turner. With Eight Illustrations in Colours and Two Woodcuts. New and Revised Edition. Crown 8vo, cloth, 3s. 6d.
Tales for the Marines. Post 8vo, illustrated boards, 2s.

Treeton (Ernest A.).—The Instigator. Cr. 8vo, cloth, gilt top, 6s.

Twain's (Mark) Books.

Author's Edition de Luxe of the Works of Mark Twain, in 22 Volumes (limited to 600 Numbered Copies), price 12s. 6d. net per Volume (Sold only in Sets.)

UNIFORM LIBRARY EDITION. Crown 8vo, cloth extra, 3s. 6d. each.

Mark Twain's Library of Humour. With 197 Illustrations by E. W. KEMBLE.
Roughing It; and The Innocents at Home. With 200 Illustrations by F. A. FRASER.
The American Claimant. With 81 Illustrations by HAL HURST and others.
*The Adventures of Tom Sawyer. With 111 Illustrations.
Tom Sawyer Abroad. With 26 Illustrations by DAN BEARD.
Tom Sawyer, Detective, &c. With Photogravure Portrait of the Author.
Pudd'nhead Wilson. With Portrait and Six Illlustrations by LOUIS LOEB.
*A Tramp Abroad. With 314 Illustrations.
*The Innocents Abroad; or, The New Pilgrim's Progress. With 234 Illustrations. (The Two Shilling Edition is entitled Mark Twain's Pleasure Trip.)
*The Gilded Age. By MARK TWAIN and C. D. WARNER With 212 Illustrations.
*The Prince and the Pauper. With 190 Illustrations.
*Life on the Mississippi. With 300 Illustrations.
*The Adventures of Huckleberry Finn. With 174 Illustrations by E. W. KEMBLE.
*A Yankee at the Court of King Arthur. With 220 Illustrations by DAN BEARD.
*The Stolen White Elephant. | *The £1,000,000 Bank-Note.
A Double-barrelled Detective Story. With 7 Illustrations by LUCIUS HITCHCOCK.
The Choice Works of Mark Twain. Revised and Corrected throughout by the Author. With Life, Portrait, and numerous Illustrations. | Christian Science.   [Shortly.

*** The books marked * may be had also in post 8vo, picture boards, at 2s. each.

Crown 8vo, cloth, gilt top, 6s. each.

Personal Recollections of Joan of Arc. With Twelve Illustrations by F. V. DU MOND.
More Tramps Abroad.
The Man that Corrupted Hadleyburg, and other Stories and Sketches. With a Frontispiece.
Mark Twain's Sketches. Pott 8vo, cloth, gilt top, 2s. net; leather, gilt edges, 3s. net.

Timbs (John), Works by. Crown 8vo, cloth, 3s. 6d. each.
  Clubs and Club Life in London: Anecdotes of its Famous Coffee-houses, Hostelries, and Taverns. With 41 Illustrations.
  English Eccentrics and Eccentricities: Stories of Delusions, Impostures, Sporting Scenes, Eccentric Artists, Theatrical Folk, &c. With 48 Illustrations.

Trollope (Anthony), Novels by.
Crown 8vo, cloth extra, 3s. 6d. each; post 8vo, illustrated boards, 2s. each.
The Way We Live Now. | Mr. Scarborough's Family.
Frau Frohmann. | Marion Fay. | The Land-Leaguers.
Post 8vo, illustrated boards, 2s. each.
Kept in the Dark. | The American Senator. | The Golden Lion of Granpere.

Trollope (Frances E.), Novels by.
Crown 8vo, cloth extra, 3s. 6d. each; post 8vo, illustrated boards, 2s. each.
Like Ships upon the Sea. | Mabel's Progress. | Anne Furness.

Trollope (T. A.).—Diamond Cut Diamond. Post 8vo, illust. bds., 2s.

Tytler (C. C. Fraser-).—Mistress Judith: A Novel. Crown 8vo, cloth extra, 3s. 6d.; post 8vo, illustrated boards, 2s.

Tytler (Sarah), Novels by.
Crown 8vo, cloth extra, 3s. 6d. each; post 8vo, illustrated boards, 2s. each.
Buried Diamonds. | The Blackhall Ghosts. | What She Came Through.
Post 8vo, illustrated boards, 2s. each.
The Bride's Pass. | The Huguenot Family. | Noblesse Oblige. | Disappeared.
Saint Mungo's City. | Lady Bell. | Beauty and the Beast.
Crown 8vo, cloth, 3s. 6d. each.
The Macdonald Lass. With Frontispiece. | Mrs. Carmichael's Goddesses.
The Witch-Wife. | Rachel Langton. | Sapphira. | A Honeymoon's Eclipse.
A Young Dragon.
Citoyenne Jacqueline. Crown 8vo, picture cloth, flat back, 2s.
Crown 8vo, cloth, gilt top, 6s. each.
Three Men of Mark. | In Clarissa's Day. | Sir David's Visit-rs.

Upward (Allen).—The Queen Against Owen. Crown 8vo, cloth, 3s. 6d.; picture cloth, flat back, 2s.; post 8vo, picture boards, 2s.

Vandam (Albert D.).—A Court Tragedy. With 6 Illustrations by J. BARNARD DAVIS. Crown 8vo, cloth, 3s. 6d.

Vashti and Esther. By 'Belle' of *The World.* Cr. 8vo, cloth, 3s. 6d.

Vizetelly (Ernest A.), Books by. Crown 8vo, cloth, 3s. 6d. each.
  The Scorpion: A Romance of Spain. With a Frontispiece. | The Lover's Progress.
  With Zola in England: A Story of Exile. With 4 Portraits.
  A Path of Thorns. Crown 8vo, cloth, gilt top, 6s.
  Bluebeard: An Account of Comorre the Cursed and Gilles de Rais; with a Summary of various Tales and Traditions. With 9 Illustrations. Demy 8vo, cloth, 9s. net.

Wagner (Leopold).—How to Get on the Stage, and how to Succeed there. Crown 8vo, cloth, 2s. 6d.

Walford's County Families of the United Kingdom (1904). Containing Notices of the Descent, Birth, Marriage, Education, &c., of more than 12,000 Distinguished Heads of Families, their Heirs Apparent or Presumptive, the Offices they hold or have held, their Town and Country Addresses, Clubs, &c. Royal 8vo, cloth gilt, 50s. *[Preparing.*

Waller (S. E.).—Sebastiani's Secret. With 9 Illusts, Cr. 8vo, cl., 6s.

Walton and Cotton's Complete Angler. With Memoirs and Notes by Sir HARRIS NICOLAS. Pott 8vo, cloth, gilt top, 2s. net; leather, gilt edges, 3s. net.

Walt Whitman, Poems by. Edited, with Introduction, by WILLIAM M. ROSSETTI. With Portrait. Crown 8vo, hand-made paper and buckram, 6s.

Warden (Florence), Novels by.
Joan, the Curate. Crown 8vo, cloth, 3s. 6d.; picture cloth, flat back, 2s.
A Fight to a Finish. Crown 8vo, cloth, 3s. 6d.
Crown 8vo, cloth, gilt top, 6s. each.
The Heart of a Girl. With 8 Illustrations. | What ought she to do? *[Shortly.*

Warman (Cy).—The Express Messenger. Crown 8vo, cloth, 3s. 6d.

Warner (Chas. Dudley).—A Roundabout Journey. Cr. 8vo, cl., 6s.

Warrant to Execute Charles I. A Facsimile, with the 59 Signatures and Seals. Printed on paper 22 in. by 14 in. 2s.
Warrant to Execute Mary Queen of Scots. A Facsimile, including Queen Elizabeth's Signature and the Great Seal. 2s.

Wassermann (Lillias).—The Daffodils. Crown 8vo, cloth, 1s. 6d.

Weather, How to Foretell the, with the Pocket Spectroscope. By F. W. CORY. With Ten Illustrations. Crown 8vo, 1s.; cloth, 1s. 6d.

Webber (Byron).—Sport and Spangles. Crown 8vo, cloth, 2s.

Werner (A.).—Chapenga's White Man. Crown 8vo, cloth, 3s. 6d.

Westbury (Atha).—The Shadow of Hilton Fernbrook: A Romance of Maoriland. Crown 8vo, cloth, 3s. 6d.

Westall (William), Novels by.
Trust Money. Crown 8vo, cloth, 3s. 6d. ; post 8vo, illustrated boards, 2s.

Crown 8vo, cloth. 6s. each.

As a Man Sows.	As Luck would have it.	The Sacred Crescents.
	The Old Bank.	

Crown 8vo, cloth 3s. 6d. each.

A Woman Tempted Him.	Nigel Fortescue.	The Phantom City.
For Honour and Life.	Ben Clough. | Birch Dene.	Ralph Norbreck's Trust.
Her Two Millions.	The Old Factory (also at 6d.)	A Queer Race.
Two Pinches of Snuff.	Sons of Belial.	Red Ryvington.
With the Red Eagle.	Strange Crimes.	Roy of Roy's Court.
A Red Bridal.	Her Ladyship's Secret.	

Wheelwright (E. Gray).—A Slow Awakening. Crown 8vo, 6s.

Whishaw (Fred.), Novels by.
A Forbidden Name: A Story of the Court of Catherine the Great. Crown 8vo, cloth, 3s. 6d.

Crown 8vo, cloth, gilt top, 6s. each.

Mazeppa.	Near the Tsar, near Death.	A Splendid Impostor.

White (Gilbert).—Natural History of Selborne. Post 8vo, 2s.

Wilde (Lady). — The Ancient Legends, Mystic Charms, and Superstitions of Ireland ; with Sketches of the Irish Past. Crown 8vo, cloth, 3s. 6d.

Williams (W. Mattieu, F.R.A.S.), Works by.
Science in Short Chapters. Crown 8vo, cloth extra, 7s. 6d.
A Simple Treatise on Heat. With Illustrations. Crown 8vo, cloth, 2s. 6d.
The Chemistry of Cookery. Crown 8vo, cloth extra, 6s.

Williamson (Mrs. F. H.).—A Child Widow. Post 8vo, bds., 2s.

Wills (C. J.), Novels by.
An Easy-going Fellow. Crown 8vo, cloth, 3s. 6d. | His Dead Past. Crown 8vo, cloth, 6s.

Wilson (Dr. Andrew, F.R.S.E.), Works by.
Chapters on Evolution. With 259 Illustrations. Crown 8vo, cloth extra 7s. 6d.
Leisure-Time Studies. With Illustrations. Crown 8vo, cloth extra, 6s.
Studies in Life and Sense. With 36 Illustrations. Crown 8vo, cloth. 3s. 6d.
Common Accidents: How to Treat Them. With Illustrations. Crown 8vo, 1s. ; cloth, 1s. 6d.
Glimpses of Nature. With 35 Illustrations. Crown 8vo, cloth extra, 3s. 6d.

Winter (John Strange), Stories by. Post 8vo, illustrated boards, 2s. each ; cloth limp, 2s. 6d. each.
Cavalry Life. | Regimental Legends.
Cavalry Life and Regimental Legends. Cr. 8vo, cloth, 3s. 6d. ; picture cloth, flat back, 2s.

Wood (H. F.), Detective Stories by. Post 8vo, boards, 2s. each.
The Passenger from Scotland Yard. | The Englishman of the Rue Cain.

Woolley (Celia Parker).—Rachel Armstrong; or, Love and Theology. Post 8vo, cloth, 2s. 6d.

Wright (Thomas, F.S.A.), Works by.
Caricature History of the Georges ; or, Annals of the House of Hanover. Compiled from Squibs, Broadsides, Window Pictures, Lampoons, and Pictorial Caricatures of the Time. With over 300 Illustrations. Crown 8vo, cloth, 3s. 6d.
History of Caricature and of the Grotesque in Art, Literature, Sculpture, and Painting. Illustrated by F. W. FAIRHOLT, F.S.A. Crown 8vo, cloth, 7s. 6d.

Wynman (Margaret).—My Flirtations. With 13 Illustrations by J. BERNARD PARTRIDGE. Post 8vo, cloth limp, 2s.

Zola's (Emile) Novels. UNIFORM EDITION. Translated or Edited, with Introductions, by ERNEST A. VIZETELLY. Crown 8vo, cloth, 3s. 6d. each.

His Masterpiece.	The Fat and the Thin.	Money.
The Joy of Life.	His Excellency.	
Germinal: Master and Man.	The Dream.	
The Honour of the Army.	The Downfall.	
Abbe Mouret's Transgression.	Doctor Pascal.	
The Fortune of the Rougons.	Lourdes.	Fruitfulness.
The Conquest of Plassans.	Rome.	Work.
The Dram-Shop.	Paris.	Truth.

POPULAR EDITIONS, medium 8vo, 6d. each.

The Dram-Shop.	The Downfall.

With Zola in England. By ERNEST A. VIZETELLY. With Four Portraits. Crown 8vo, cloth, 3s. 6d.

'ZZ' (L. Zangwill).—A Nineteenth Century Miracle. Cr. 8vo, 3s. 6d.

SOME BOOKS CLASSIFIED IN SERIES.

The St. Martin's Library. Pott 8vo, cloth, 2s. net each; leather, 3s. net each

The Woman in White. By WILKIE COLLINS.
All Sorts and Conditions of Men. By Sir WALTER BESANT.
The Cloister and the Hearth. By CHAS. READE. | 'It is Never Too Late to Mend.' By CH. READE.
Familiar Studies of Men and Books. By ROBERT LOUIS STEVENSON.
Virginibus Puerisque, and other Papers. By R. LOUIS STEVENSON.
The Pocket R.L.S.: Favourite Passages from STEVENSON'S Works.
New Arabian Nights. By ROBERT LOUIS STEVENSON. | The Deemster. By HALL CAINE.
Under the Greenwood Tree. By THOMAS HARDY. | The Life of the Fields. By RICHARD JEFFERIES.
Walton and Cotton's Complete Angler. | Mark Twain's Sketches.
Condensed Novels. (The Two Series in One Volume.) By BRET HARTE.

The Mayfair Library. Post 8vo, cloth limp, 2s. 6d. per Volume.

Quips and Quiddities. By W. D. ADAMS.
The Agony Column of 'The Times.'
A Journey Round My Room. By X. DE MAISTRE.
Poetical Ingenuities. By W. T. DOBSON.
The Cupboard Papers. By FIN-BEC.
Songs of Irish Wit and Humour.
Animals and their Masters. By Sir A HELPS.
Social Pressure. By Sir A. HELPS.
Autocrat of Breakfast-Table. By O. W. HOLMES.
Curiosities of Criticism. By H. J. JENNINGS.
Pencil and Palette. By R. KEMPT.

Little Essays: from LAMB'S LETTERS.
Forensic Anecdotes. By JACOB LARWOOD.
Theatrical Anecdotes. By JACOB LARWOOD.
Ourselves. By E. LYNN LINTON.
Witch Stories. By E. LYNN LINTON.
Pastimes and Players. By R. MACGREGOR.
New Paul and Virginia. By W. H. MALLOCK.
Puck on Pegasus. By H. C. PENNELL.
Pegasus Re-saddled. By H. C. PENNELL.
The Muses of Mayfair. By H. C. PENNELL.
By Stream and Sea. By WILLIAM SENIOR.

The Golden Library. Post 8vo, cloth limp, 2s. per Volume.

Songs for Sailors. By W. C. BENNETT.
Lives of the Necromancers. By W. GODWIN.
The Autocrat of the Breakfast Table. By OLIVER WENDELL HOLMES.

Scenes of Country Life. By EDWARD JESSE.
La Mort d'Arthur: Selections from MALLORY.
The Poetical Works of Alexander Pope.
Diversions of the Echo Club. BAYARD TAYLOR.

My Library. Printed on laid paper, post 8vo, half-Roxburghe, 2s. 6d. each.

The Journal of Maurice de Guerin.
The Dramatic Essays of Charles Lamb.
Citation of William Shakspeare. W. S. LANDOR.

Christie Johnstone. By CHARLES READE.
Peg Woffington. By CHARLES READE.

The Pocket Library. Post 8vo, printed on laid paper and hf.-bd., 2s. each.

Gastronomy. By BRILLAT-SAVARIN.
Robinson Crusoe. Illustrated by G. CRUIKSHANK
Autocrat and Professor. By O. W. HOLMES.
Provincial Letters of Blaise Pascal.
Whims and Oddities. By THOMAS HOOD.
Leigh Hunt's Essays. Edited by E. OLLIER.
The Barber's Chair. By DOUGLAS JERROLD.

The Essays of Elia. By CHARLES LAMB.
Anecdotes of the Clergy. By JACOB LARWOOD.
The Epicurean, &c. By THOMAS MOORE.
Plays by RICHARD BRINSLEY SHERIDAN.
Gulliver's Travels, &c. By Dean SWIFT.
Thomson's Seasons. Illustrated.
White's Natural History of Selborne.

POPULAR SIXPENNY NOVELS.

The Tents of Shem By GRANT ALLEN.
The Orange Girl. By WALTER BESANT.
All Sorts and Conditions of Men. WALT. BESANT.
Children of Gibeon. By WALTER BESANT.
The Chaplain of the Fleet. BESANT and RICE.
Ready-Money Mortiboy. BESANT and RICE.
The Golden Butterfly. BESANT and RICE.
Shadow of the Sword. By R. BUCHANAN.
The Deemster. By HALL CAINE.
The Shadow of a Crime. By HALL CAINE.
A Son of Hagar. By HALL CAINE.
Antonina. By WILKIE COLLINS.
Armadale. By WILKIE COLLINS.
The Moonstone. By WILKIE COLLINS.
The Woman in White. By WILKIE COLLINS.
The Dead Secret. By WILKIE COLLINS.
Man and Wife. By WILKIE COLLINS.
The New Magdalen. By WILKIE COLLINS.

Diana Barrington. By B. M. CROKER.
Joseph's Coat. By D. CHRISTIE MURRAY.
Held in Bondage. By OUIDA.
Moths. By OUIDA.
Puck. By OUIDA.
Under Two Flags. By OUIDA.
Strathmore. By OUIDA.
Peg Woffington; and Christie Johnstone. By CHARLES READE.
Cloister and the Hearth. By CHARLES READE.
Griffith Gaunt. By CHARLES READE.
It is Never Too Late to Mend. CHARLES READE.
Hard Cash. By CHARLES READE.
Foul Play. By CHARLES READE.
New Arabian Nights. By R. L. STEVENSON.
The Old Factory. By WILLIAM WESTALL.
The Downfall. By EMILE ZOLA.
The Dram-Shop. By EMILE ZOLA.

THE PICCADILLY NOVELS.

LIBRARY EDITIONS OF NOVELS, many Illustrated, crown 8vo, cloth extra, 3s. 6d. each.

By Mrs. ALEXANDER.

Valerie's Fate.
A Life Interest.
Mona's Choice.
By Woman's Wit.
The Cost of Her Pride.
Barbara.
A Fight with Fate.
A Golden Autumn.
Mrs Crichton's Creditor.
The Step-mother.
A Missing Hero.

By M. ANDERSON.—Othello's Occupation.

By G. WEBB APPLETON.
Rash Conclusions.

By EDWIN L. ARNOLD.
Phra the Phoenician. | Constable of St. Nicholas.

By ARTEMUS WARD
Artemus Ward Complete.

By F. M. ALLEN.—Green as Grass.

By GRANT ALLEN.

Philistia. | Babylon.
Strange Stories.
For Maimie's Sake.
In all Shades.
The Beckoning Hand.
The Devil's Die.
This Mortal Coil.
The Tents of Shem.
The Great Taboo.
Dumaresq's Daughter.
Duchess of Powysland.
Blood Royal.
I. Greet's Masterpiece.
The Scallywag.
At Market Value.
Under Sealed Orders.

By ROBERT BARR
In a Steamer Chair.
From Whose Bourne.
A Woman Intervenes.
Revenge!

THE PICCADILLY (3/6) NOVELS—*continued.*

By FRANK BARRETT.

Woman of Iron Bracelets. | Under a Strange Mask.
Fettered for Life. | A Missing Witness.
The Harding Scandal. | Was She Justified?

By 'BELLE.'—Vashti and Esther.

By ARNOLD BENNETT.

The Gates of Wrath.

By Sir W. BESANT and J. RICE.

Ready-Money Mortiboy. | By Celia's Arbour.
My Little Girl. | Chaplain of the Fleet.
With Harp and Crown. | The Seamy Side.
This Son of Vulcan. | The Case of Mr. Lucraft.
The Golden Butterfly. | In Trafalgar's Bay.
The Monks of Thelema. | The Ten Years' Tenant.

By Sir WALTER BESANT.

All Sorts & Conditions. | Armorel of Lyonesse.
The Captains' Room. | S. Katherine's by Tower
All in a Garden Fair. | Verbena Camellia, &c.
Dorothy Forster. | The Ivory Gate.
Uncle Jack. | Holy Rose | The Rebel Queen.
World Went Well Then. | Dreams of Avarice.
Children of Gibeon. | In Deacon's Orders.
Herr Paulus. | The Master Craftsman.
For Faith and Freedom. | The City of Refuge.
To Call Her Mine. | A Fountain Sealed.
The Revolt of Man. | The Changeling.
The Bell of St. Paul's. | The Fourth Generation
The Charm.

By AMBROSE BIERCE—In Midst of Life.
By HAROLD BINDLOSS. Ainslie's Ju-Ju.

By M. McD. BODKIN.

Dora Myrl. | Shillelagh and Shamrock.

By PAUL BOURGET.—A Living Lie.
By J. D. BRAYSHAW.—Slum Silhouettes.
By H. A. BRYDEN.—An Exiled Scot.

By ROBERT BUCHANAN.

Shadow of the Sword. | The New Abelard.
A Child of Nature. | Matt. | Rachel Dene
God and the Man. | Master of the Mine.
Martyrdom of Madeline | The Heir of Linne.
Love Me for Ever. | Woman and the Man.
Annan Water. | Red and White Heather.
Foxglove Manor. | Lady Kilpatrick.
The Charlatan. | Andromeda.

By GELETT BURGESS and WILL
IRWIN.—The Picaroons.
R. W. CHAMBERS.—The King in Yellow.
By J. M. CHAPPLE.—The Minor Chord.

By HALL CAINE.

Shadow of a Crime. | Deemster. | Son of Hagar.

By AUSTIN CLARE.—By Rise of River.

By Mrs. ARCHER CLIVE.

Paul Ferroll. | Why Paul Ferroll Killed his Wife.

By ANNE COATES.—Rie's Diary.

By MACLAREN COBBAN.

The Red Sultan. | The Burden of Isabel.

By WILKIE COLLINS.

Armadale. | After Dark. | The New Magdalen.
No Name. | Antonina | The Frozen Deep.
Basil. | Hide and Seek. | The Two Destinies.
The Dead Secret. | 'I Say No.'
Queen of Hearts. | Little Novels.
My Miscellanies. | The Fallen Leaves.
The Woman in White. | Jezebel's Daughter.
The Law and the Lady. | The Black Robe.
The Haunted Hotel. | Heart and Science.
The Moonstone. | The Evil Genius.
Man and Wife. | The Legacy of Cain.
Poor Miss Finch. | A Rogue's Life.
Miss or Mrs.? | Blind Love.

By MORT. & FRANCES COLLINS.

Blacksmith & Scholar. | You Play me False.
The Village Comedy. | Midnight to Midnight.

M. J. COLQUHOUN.—Every Inch Soldier.

By HERBERT COMPTON.

The Inimitable Mrs. Massingham.

By E. H. COOPER.—Geoffory Hamilton.
By V. C. COTES.—Two Girls on a Barge.

By C. E. CRADDOCK.

The Prophet of the Great Smoky Mountains.
His Vanished Star.

By H. N. CRELLIN.

Romances of the Old Seraglio.

By MATT CRIM.

The Adventures of a Fair Rebel.

By S. R. CROCKETT and others.

Tales of Our Coast.

By B. M. CROKER.

Diana Barrington. | The Real Lady Hilda.
Proper Pride. | Married or Single?
A Family Likeness. | Two Masters.
Pretty Miss Neville. | In the Kingdom of Kerry
A Bird of Passage. | Interference.
Mr. Jervis. | A Third Person.
Village Tales. | Beyond the Pale.
Some One Else. | Jason. | Miss Balmaine's Past.
Infatuation. | Terence.

By ALPHONSE DAUDET.

The Evangelist ; or, Port Salvation.

H. C. DAVIDSON.—Mr. Sadler's Daughters.

By JAS. DE MILLE.

A Strange Manuscript Found in a Copper Cylinder.

By HARRY DE WINDT.

True Tales of Travel and Adventure.

By DICK DONOVAN.

Man from Manchester. | Tales of Terror.
Records of Vincent Trill | Chronicles of Michael
The Mystery of | Danevitch. [Detective.
Jamaica Terrace. | Tyler Tatlock, Private
Deacon Brodie.

By RICHARD DOWLING.

Old Corcoran's Money.

By A. CONAN DOYLE.

The Firm of Girdlestone.

By S. JEANNETTE DUNCAN.

A Daughter of To-day. | Vernon's Aunt.

By ANNIE EDWARDES.

Archie Lovell. | A Plaster Saint.

By G. S. EDWARDS.—Snazelleparilla.

By G. MANVILLE FENN

Cursed by a Fortune. | A Fluttered Dovecote.
The Case of Ailsa Gray | King of the Castle
Commodore Junk. | Master of Ceremonies.
The New Mistress. | The Man with a Shadow
Witness to the Deed. | One Maid's Mischief.
The Tiger Lily. | Story of Antony Grace.
The White Virgin. | This Man's Wife.
Black Blood. | In Jeopardy.   [ning
Double Cunning. | A Woman Worth Win-

By PERCY FITZGERALD.—Fatal Zero
By Hon. Mrs. W. FORBES.—Dumb.

By R. E. FRANCILLON.

One by One. | Ropes of Sand.
A Dog and his Shadow. | Jack Doyle's Daughter.
A Real Queen.

By HAROLD FREDERIC.

Seth's Brother's Wife. | The Lawton Girl.

By PAUL GAULOT.—The Red Shirts.

By CHARLES GIBBON.

Robin Gray. | The Golden Shaft.
Loving a Dream. | The Braes of Yarrow.
Of High Degree | Queen of the Meadow.
The Flower of the Forest.

By E. GLANVILLE.

The Lost Heiress. | The Golden Rock.
Fair Colonist | Fossicker | Tales from the Veld.

By E. J. GOODMAN.

The Fate of Herbert Wayne.

By Rev. S. BARING GOULD.

Red Spider. | Eve.

By ALFRED A. GRACE.

Tales of a Dying Race.

CECIL GRIFFITH.—Corinthia Marazion.

By A. CLAVERING GUNTER.

A Florida Enchantment.

By BRET HARTE.

A Waif of the Plains. | A Protégée of Jack
A Ward of the Golden | Clarence.   [Hamlin's
Gate.   (Springs. | Barker's Luck.
A Sappho of Green | Devil's Ford.   [celsior.'
Col. Starbottle's Client. | The Crusade of the 'Ex-
Sasy. | Sally Dows. | Three Partners.
Bell-Ringer of Angel's. | Gabriel Conroy.
Tales of Trail and Town | New Condensed Novels

THE PICCADILLY (3/6) NOVELS—*continued*.

By AMELIE RIVES.
Barbara Dering. | Meriel.

By F. W. ROBINSON.
The Hands of Justice. | Woman in the Dark.

By ALBERT ROSS.—A Sugar Princess.

J. RUNCIMAN.—Skippers and Shellbacks.

By W. CLARK RUSSELL.
Round the Galley-Fire. | My Shipmate Louise.
In the Middle Watch. | Alone on Wide Wide Sea.
On the Fo'k'sle Head | The Phantom Death.
A Voyage to the Cape. | Is He the Man?
Book for the Hammock. | Good Ship 'Mohock.'
Mystery of 'Ocean Star' | The Convict Ship.
Jenny Harlowe. | Heart of Oak.
An Ocean Tragedy. | The Tale of the Ten.
A Tale of Two Tunnels. | The Last Entry.
The Death Ship.

By DORA RUSSELL.—Drift of Fate.

By HERBERT RUSSELL.   True Blue.

BAYLE ST. JOHN.—A Levantine Family.

By ADELINE SERGEANT.
Dr. Endicott's Experiment.
Under False Pretences.

By M. P. SHIEL.—The Purple Cloud.

By GEORGE R. SIMS.
Dagonet Abroad. | In London's Heart.
Once Upon a Christmas | Mary Jane's Memoirs.
Time. | Mary Jane Married.
Without the Limelight. | The Small-part Lady.
Rogues and Vagabonds. | A Blind Marriage.
Biographs of Babylon.

By UPTON SINCLAIR.—Prince Hagen.

By HAWLEY SMART.
Without Love or Licence. | The Outsider.
The Master of Rathkelly. | Beatrice & Benedick.
Long Odds. | A Racing Rubber.

By J. MOYR SMITH.
The Prince of Argolis.

By T. W. SPEIGHT.
A Secret of the Sea. | A Minion of the Moon.
The Grey Monk. | Secret Wyvern Towers.
The Master of Trenance | The Doom of Siva.
The Web of Fate. | As it was Written.
The Strange Experiences of Mr. Verschoyle.
Her Ladyship.

By ALAN ST. AUBYN.
A Fellow of Trinity. | The Tremlett Diamonds.
The Junior Dean. | The Wooing of May
Master of St. Benedict's. | A Tragic Honeymoon.
To his Own Master. | A Proctor's Wooing.
Gallantry Bower. | Fortune's Gate.
In Face of the World. | Bonnie Maggie Lauder.
Orchard Damerel. | Mary Unwin.
Mrs. Dunbar's Secret.

By JOHN STAFFORD.—Doris and I.

By R. STEPHENS.—The Cruciform Mark.

By R. NEILSON STEPHENS.
Philip Winwood.

R. A. STERNDALE.—The Afghan Knife.

R. L. STEVENSON.—The Suicide Club.

By FRANK STOCKTON.
The Young Master of Hyson Hall.

By SUNDOWNER.   Told by the Taffrail.

By ANNIE THOMAS.—The Siren's Web.

By BERTHA THOMAS.
The Violin-Player. | In a Cathedral City.

By FRANCES E. TROLLOPE
Like Ships upon Sea. | Mabel's Progress.
Anne Furness.

By ANTHONY TROLLOPE.
The Way we Live Now. | Scarborough's Family.
Frau Frohmann. | The Land-Leaguers.
Marion Fay.

By IVAN TURGENIEFF, &c.
Stories from Foreign Novelists.

By MARK TWAIN.
Choice Works. | Pudd'nhead Wilson.
Library of Humour. | The Gilded Age.
The Innocents Abroad. | Prince and the Pauper.
Roughing It; and The | Life on the Mississippi.
Innocents at Home. | The Adventures of
A Tramp Abroad. | Huckleberry Finn.
The American Claimant. | A Yankee at the Court
Adventures Tom Sawyer | of King Arthur.
Tom Sawyer Abroad. | Stolen White Elephant.
Tom Sawyer, Detective | £1,000,000 Bank-note.
A Double-barrelled Detective Story.

C. C. F.-TYTLER.—Mistress Judith.

By SARAH TYTLER.
What She Came Through | Mrs. Carmichael's God-
Buried Diamonds. | desses.
The Blackhall Ghosts | Rachel Langton.
The Macdonald Lass. | A Honeymoon's Eclipse.
Witch-Wife. | Sapphira | A Young Dragon.

By ALLEN UPWARD.
The Queen against Owen.

By ALBERT D. VANDAM.
A Court Tragedy.

By E. A. VIZETELLY.
The Scorpion. | The Lover's Progress.

By FLORENCE WARDEN.
Joan, the Curate. | A Fight to a Finish.

By CY WARMAN.—Express Messenger.

By A. WERNER.
Chapenga's White Man.

By WILLIAM WESTALL.
For Honour and Life. | The Old Factory.
A Woman Tempted Him | Red Ryvington.
Her Two Millions. | Ra'ph Norbreck's Trust
Two Pinches of Snuff. | Trust-money.
Nigel Fortescue. | Sons of Belial.
Birch Dene. | Roy of Roy's Court.
The Phantom City. | With the Red Eagle.
A Queer Race. | A Red Bridal.
Ben Clough. | Strange Crimes.
Her Ladyship's Secret.

By ATHA WESTBURY.
The Shadow of Hilton Fernbrook.

By FRED WHISHAW.
A Forbidden Name.

By C. J. WILLS.—An Easy-going Fellow.

By JOHN STRANGE WINTER.
Cavalry Life; and Regimental Legends.

By E. ZOLA.
The Joy of Life. | His Masterpiece.
The Fortune of the Rougons.
Abbe Mouret's Transgression.
The Conquest of Plassans. | Germinal.
The Honour of the Army.
The Downfall. | His Excellency.
The Dream. | Money. | The Dram Shop.
Dr. Pascal. | Lourdes. | Rome. | Paris. | Work.
The Fat and the Thin. | Fruitfulness. | Truth.

By 'ZZ.'—A Nineteenth Century Miracle.

CHEAP EDITIONS OF POPULAR NOVELS.
Post 8vo, illustrated boards, 2s. each.

By ARTEMUS WARD.
Artemus Ward Complete.

By Mrs. ALEXANDER.
Maid, Wife, or Widow? | A Life Interest.
Blind Fate. | Mona's Choice.
Valerie's Fate. | By Woman's Wit.

By E. LESTER ARNOLD.
Phra the Phœnician.

By GRANT ALLEN.
Philistia. | Babylon. | Dumaresq's Daughter.
Strange Stories. | Duchess of Powysland.
For Maimie's Sake. | Blood Royal. [piece.
In all Shades. | Ivan Greet's Master-
The Beckoning Hand. | The Scallywag.
The Devil's Die. | This Mortal Coil.
The Tents of Shem | At Market Value.
The Great Taboo. | Under Sealed Orders.

TWO-SHILLING NOVELS—*continued*.

BY FRANK BARRETT.

Fettered for Life.	Found Guilty.
Little Lady Linton.	A Recoiling Vengeance.
Between Life & Death.	For Love and Honour.
Sin of Olga Zassoulich.	John Ford, &c.
Folly Morrison.	Woman of Iron Brace'ts
Lieut. Barnabas.	The Harding Scandal.
Honest Davie.	A Missing Witness.
A Prodigal's Progress.	

By Sir W. BESANT and J. RICE.

Ready-Money Mortiboy.	By Celia's Arbour.
My Little Girl.	Chaplain of the Fleet.
With Harp and Crown.	The Seamy Side.
This Son of Vulcan.	The Case of Mr. Lucraft.
The Golden Butterfly.	In Trafalgar's Bay.
The Monks of Thelema.	The Ten Years' Tenant.

By Sir WALTER BESANT.

All Sorts and Conditions of Men.	The Bell of St. Paul's.
The Captains' Room.	The Holy Rose.
All in a Garden Fair.	Armorel of Lyonesse.
Dorothy Forster.	S. Katherine's by Tower
Uncle Jack.	Verbena Camellia Stephanotis.
The World Went Very Well Then.	The Ivory Gate.
Children of Gibeon.	The Rebel Queen.
Herr Paulus.	Beyond the Dreams of Avarice.
For Faith and Freedom.	The Revolt of Man.
To Call Her Mine.	In Deacon's Orders.
The Master Craftsman.	The City of Refuge.

By AMBROSE BIERCE.

In the Midst of Life.

By FREDERICK BOYLE.

Camp Notes.	Chronicles of No-man's Land.
Savage Life.	

BY BRET HARTE.

Californian Stories.	Flip. Maruja.
Gabriel Conroy.	A Phyllis of the Sierras.
Luck of Roaring Camp.	A Waif of the Plains.
An Heiress of Red Dog.	Ward of Golden Gate.

By ROBERT BUCHANAN.

Shadow of the Sword.	The Martyrdom of Madeline.
A Child of Nature.	The New Abelard.
God and the Man.	The Heir of Linne.
Love Me for Ever.	Woman and the Man.
Foxglove Manor.	Rachel Dene. Matt.
The Master of the Mine.	Lady Kilpatrick.
Annan Water.	

By BUCHANAN and MURRAY.

The Charlatan.

By HALL CAINE.

The Shadow of a Crime.	The Deemster.
A Son of Hagar.	

By Commander CAMERON.

The Cruise of the 'Black Prince.'

By HAYDEN CARRUTH.

The Adventures of Jones.

By AUSTIN CLARE.

For the Love of a Lass.

By Mrs. ARCHER CLIVE.

Paul Ferroll.
Why Paul Ferroll Killed his Wife.

By MACLAREN COBBAN.

The Cure of Souls.	The Red Sultan.

By WILKIE COLLINS.

Armadale. After Dark.	My Miscellanies.
No Name.	The Woman in White.
Antonina.	The Moonstone.
Basil.	Man and Wife.
Hide and Seek.	Poor Miss Finch.
The Dead Secret.	The Fallen Leaves.
Queen of Hearts.	Jezebel's Daughter.
Miss or Mrs.?	The Black Robe.
The New Magdalen.	Heart and Science.
The Frozen Deep.	'I Say No!'
The Law and the Lady	The Evil Genius.
The Two Destinies.	Little Novels.
The Haunted Hotel.	Legacy of Cain.
A Rogue's Life.	Blind Love.

By C. ALLSTON COLLINS.

The Bar Sinister.

By MORT. & FRANCES COLLINS

Sweet Anne Page.	Sweet and Twenty.
Transmigration.	The Village Comedy.
From Midnight to Midnight.	You Play me False.
	Blacksmith and Scholar
A Fight with Fortune.	Frances.

By M. J. COLQUHOUN.

Every Inch a Soldier.

By C. EGBERT CRADDOCK.

The Prophet of the Great Smoky Mountains.

By MATT CRIM.

The Adventures of a Fair Rebel.

By H. N. CRELLIN.—Tales of the Caliph.

By B. M. CROKER.

Pretty Miss Neville.	Village Tales and Jungle Tragedies.
Diana Barrington.	
'To Let.'	Two Masters.
A Bird of Passage.	Mr. Jervis.
Proper Pride.	The Real Lady Hilda.
A Family Likeness.	Married or Single?
A Third Person.	Interference.

By ALPHONSE DAUDET.

The Evangelist; or, Port Salvation.

By JAMES DE MILLE.

A Strange Manuscript.

By DICK DONOVAN.

The Man-Hunter.	In the Grip of the Law.
Tracked and Taken.	From Information Received.
Caught at Last!	
Wanted!	Tracked to Doom.
Who Poisoned Hetty Duncan?	Link by Link
	Suspicion Aroused.
Man from Manchester.	Riddles Read.
A Detective's Triumphs	
The Mystery of Jamaica Terrace.	
The Chronicles of Michael Danevitch.	

By Mrs. ANNIE EDWARDES.

A Point of Honour.	Archie Lovell.

By EDWARD EGGLESTON.

Roxy.

By G. MANVILLE FENN.

The New Mistress.	The Tiger Lily.
Witness to the Deed.	The White Virgin.

By PERCY FITZGERALD.

Bella Donna.	Second Mrs. Tillotson.
Never Forgotten.	Seventy-five Brooke Street.
Polly.	
Fatal Zero.	The Lady of Brantome

By P. FITZGERALD and others.

Strange Secrets.

By R. E. FRANCILLON.

Olympia.	King or Knave?
One by One.	Romances of the Law.
A Real Queen.	Ropes of Sand.
Queen Cophetua.	A Dog and his Shadow

By HAROLD FREDERIC.

Seth's Brother's Wife.	The Lawton Girl.

Prefaced by Sir BARTLE FRERE.

Pandurang Hari.

By CHARLES GIBBON.

Robin Gray.	In Honour Bound.
Fancy Free.	Flower of the Forest.
For Lack of Gold.	The Braes of Yarrow.
What will World Say?	The Golden Shaft.
In Love and War.	Of High Degree.
For the King.	By Mead and Stream.
In Pastures Green.	Loving a Dream.
Queen of the Meadow.	A Hard Knot.
A Heart's Problem.	Heart's Delight.
The Dead Heart.	Blood-Money.

By WILLIAM GILBERT.

James Duke.

By ERNEST GLANVILLE

The Lost Heiress.	The Fossicker.
A Fair Colonist.	

By Rev. S. BARING GOULD

Red Spider.	Eve.

www.ingramcontent.com/pod-product-compliance
Lightning Source LLC
Chambersburg PA
CBHW031147120726
47905CB00006B/1846